Art Encounters in the Nuclear Age

Written by artists, curators, historians, and cultural workers, *Art Encounters in the Nuclear Age: Radiant Objects* is a collection of essays that offers an idiosyncratic cross-section of nuclear history, foregrounding stories of objects and artworks.

This collection aims to expand discourse around nuclear power and bear witness to the strangeness of living in the nuclear age and to contribute to a culture of care for nuclear subjects. Particularly the collection seeks to direct attention to the marginalised voices of those who have experienced nuclear traumas more directly. The book can be read as a kind of inventory of objects, of things close-to-hand, which can allow us to sense the magnitude and complexity of issues arising from nuclear technologies. Amidst prevailing eco-anxiety, these essays ask us to sit with these experiences, to 'hold' these objects and to come closer to understanding our nuclear heritage. Topics include: the legacies of nuclear bombs; the nuclear archive and 'noise prints', nuclear decommissioning; the past and present of nuclear bunkers; and art in Fukushima.

This text is suitable for researchers in the visual arts, history, philosophy, and environmentalism.

Jessica Holtaway is Lecturer at Solent University, UK.

Routledge Focus on Art History and Visual Studies

Routledge Focus on Art History and Visual Studies presents short-form books on varied topics within the fields of art history and visual studies.

Bodily Engagements with Film, Images, and Technology
Somavision
Max Ryynänen

Performance, Art and Politics in the African Diaspora
Necropolitics and the Black Body
Myron M. Beasley

Contemporary Art, Systems, and the Aesthetics of Dispersion
Francis Halsall

A Philosophy of Cultural Scenes in Art and Popular Culture
Max Ryynänen and Jozef Kovalčik

Translation and Transgression in the Art of Shirin Neshat
Erin C. Devine

Maya Lin, Public Art, and the Confluence Project
Matthew Reynolds

Concerning the Quantum in Art
Paul Thomas

Art Encounters in the Nuclear Age
Radiant Objects
Edited by Jessica Holtaway

For more information about this series, please visit: https://www.routledge.com/Routledge-Focus-on-Art-History-and-Visual-Studies/book-series/FOCUSAH

Art Encounters in the Nuclear Age

Radiant Objects

Edited by Jessica Holtaway

Routledge
Taylor & Francis Group
NEW YORK AND LONDON

First published 2026
by Routledge
605 Third Avenue, New York, NY 10158

and by Routledge
4 Park Square, Milton Park, Abingdon, Oxon, OX14 4RN

Routledge is an imprint of the Taylor & Francis Group, an informa business

ISBN: 978-1-041-03084-3 (hbk)
ISBN: 978-1-041-03085-0 (pbk)
ISBN: 978-1-003-62218-5 (ebk)

DOI: 10.4324/9781003622185

Typeset in Times New Roman
by KnowledgeWorks Global Ltd.

Contents

Figures

Contributors

Tim Andreae is an artist, musician, and therapist. He is a founder of the educational non-profit and eco-cultural residency, polum bolum (https://www.polumbolum.org/) and started *Holding What Can't Be Held*, a project that brings artists' perspectives to bear on the Idaho National Laboratory and its radioactive legacy http://thefeelingbody.org/.

Daniel Beck is a sound and film practitioner based in the UK and France. In 2024, he completed a PhD at the University of the Arts London, titled 'Post-atomic Auralities: A practice-based study of sound and atmospheric effects'. His practice considers how sound and atmospheric conditions can interact within sites of nuclear significance.

Jonathan Faiers is Professor of Fashion Thinking, University of Southampton, UK, where his research examines the interface between popular culture, fashion, and textiles. Jonathan's recent projects include the curation of the exhibition *Tartan* for V&A Dundee (2023–2024) and his forthcoming book, *The Sound of Fashion Thinking* (2025), Sternberg Press.

Dimitra Gkitsa is Lecturer in Curating and Cultural Leadership at the Winchester School of Art, University of Southampton. Her research and curatorial practice are situated in the intersection of memory studies, affect, activism, contemporary art and curatorial practices, and the post-socialist visual cultures.

Helen Grove-White holds a PhD from The Courtauld Institute of Art (1980) and a Fine Art degree from Lancaster (2005). She initiated the Power in the Land project (2014). Her multimedia practice explores her relationship to the land, time, archaeological remains, the legacy of nuclear power, and energy futures.

Marc Herbst is a research professor at the Free University of Bozen-Bolzano. He has written previously about purposefully progressive developments made possible by the criminal economy that flourished in the cannabis growing regions of northern California.

Jessica Holtaway is Lecturer at Southampton Solent University. Her research often centres around curatorial practice, including the 2023 exhibition *Radiant Objects: Encounters in the Nuclear Age*. Jessica is interested in how art can help us think about our relationship to our environment in a new and potentially radical ways, to foster a culture of care.

Atsuhide Ito is Senior Research Fellow at Southampton Solent University and Associate Lecturer at London College of Communication, University of the Arts London. Atsuhide's research concerns aeropolitics and discourses of nuclearity, and his practice combines environmental data exploration with endurance running.

Yoi Kawakubo is a Japanese artist based in London and Tokyo, and currently a PhD candidate at Southampton Solent University. His work explores the limits of knowledge through photography, film, sound, scent-based painting, garden design, and neon, often engaging with financial systems, nuclear energy, and both personal and historical narratives.

Bridget Kennedy is a member of the Nuclear Culture Research Group, currently undertaking a PhD in Fine Art (Goldsmiths, UAL), using a process-based, sculptural practice to explore correlatives between the materiality and social rituals of the nuclear industrial landscape of Sellafield, West Cumbria.

Min-Kyoo Kim is a PhD candidate in Film and Screen Studies at the University of Cambridge. His doctoral thesis explores visual representations of nuclear violence and victimhood, with a particular focus on those occluded and forgotten histories that are otherwise marginalised in the archives.

Michael Sanders is a freelance artist. His practice subverts the everyday, sometimes using textiles and performance to challenge and re-appropriate the imagery and symbolism of military power and technology. He exhibits internationally and is winner of the OPEM 2011 Commission Prize. A sample of his adaptation of *Polaris Military* tartan made up in nuclear text is in the collection of National Museums Scotland.

Robin Tarbet studied at the Royal College of Art and is a member of the Royal Society of Sculptors in London. Robin's work uses print and ad-hoc casting processes to make sculpture that plays with the transfer of information from discarded technologies to the edge of the natural world, with a speculative lure for perishable subject matter and systematic failure.

Jason Waite is a curator, writer, and cultural worker focused on forms of practice producing agency, currently teaching Art History at St. Anne's College, University of Oxford. He is part of the collective, Don't Follow the Wind, and works in sites of crisis amidst the detritus of capitalism, looking for tools and radical imaginaries for different ways of living and working together.

Acknowledgements

It has been a privilege to extend the curation of *Radiant Objects: Encounters in the Nuclear Age* (2023) into this edited collection, continuing and expanding the conversations between so many dedicated artists and writers. I am deeply grateful to the Fringe Arts Bath team and the participating artists whose contributions shaped the original exhibition and helped bring this book to life.

I owe particular thanks to the Nuclear Culture Research Group – without this vital network, neither the exhibition nor the book would likely have been realised. My thanks also go to Kaori Homma and the Art in the Nuclear Age (AiNA) hub. The AiNA reading group has been invaluable in thinking through the core themes explored in this collection.

I am especially grateful to Dr Jon Trayner for his attentive reading and thoughtful feedback, and to the editorial team at Routledge for their guidance and support throughout the publishing process.

A heartfelt thank you to my partner, Oliver Holtaway, for his constant encouragement and support during every stage of this project. Finally, I would like to thank my friends for their generosity and good humour – for watching documentaries with me and engaging in discussions about nuclearity.

Introduction

Jessica Holtaway

In 2016, I attended a screening of the documentary *Nobody Told Us Anything*, directed by Charles Stewart (2014). The documentary tells the stories of the men and women who participated in the British Nuclear Weapons Tests that took place between 1952 and 1967. Through interviews with veterans and their families, combined with archive footage, the film raises awareness of the physical impacts of the tests and the ethical questions surrounding them.

The screening was followed by a Q&A with the director. From memory, about fifteen people attended this event – mostly test veterans, accompanied by one or two family members. The film was impactful, and, afterwards, one of the veterans stood up and addressed the audience. I forget his exact words, but I clearly remember that he held up a sunhat that he had worn on the day of one of the nuclear tests. The hat was full of holes caused by its exposure to radioactivity, to the extent that it almost resembled crude lacework. The hat's owner was elderly: he was surviving the impacts of the radiation when many of his companions from military service had not. He presented a range of emotions of which humour prevailed, albeit a kind of enraged, determined humour in the face of trauma, perhaps a sense of the absurdity, the wildly unreasonable nature, of what he had lived through.

This image of the holey hat – a hat still radiating with nuclear contamination, held aloft before the bright cinema screen for the attention of the few people gathered in this space – left a strong impression on me. In this moment, I experienced a vertiginous sense of the scale and pervasiveness of nuclearism and the ways in which it plays out through human hierarchies of power. This manifests specifically through the ways in which different political systems have been complicit in concealing or blocking public understanding of the risks and impacts of nuclear technologies. The encounter with the holey hat gave me a heightened awareness of my own position as a nuclear subject, in no way equivalent to the owner of the hat, but as someone also entangled in the weirdness of this nuclear era.

This book can be read as a kind of inventory of objects, of things close-to-hand, that can allow us to sense the magnitude and complexity of issues arising from nuclear contamination. Amidst prevailing eco-anxiety, these

DOI: 10.4324/9781003622185-1

essays ask us to sit with these experiences, to 'hold' these objects and to come closer to understanding our reality, even as it shifts beyond our abilities of comprehension. This collection aims to bear witness to the strangeness of living in the nuclear age and to contribute to a culture of care for nuclear subjects. This means all of us, but particularly it aims to direct attention to the hushed, marginalised voices of those who have experienced nuclear traumas more directly. These voices articulate, but, at the same time, rise above loss and sadness. They often resonate with kindness, humour, and hope. This collection is therefore an idiosyncratic cross-section of nuclear history, inviting us to look at histories of the nuclear through stories of objects and artworks.

The essays contained in this book developed from an exhibition I curated in Bath, UK, in 2023, entitled *Radiant Objects: Encounters in the Nuclear Age*. Some of the contributors to this book also shared work in the exhibition, and others joined discussions and shared ideas and research following it. The open call-out for the exhibition (which was part of the Fringe Arts Bath festival) was shared with the Nuclear Culture Research Group, a global and interdisciplinary group of artists, curators and scholars, established in 2011 by Dr Ele Carpenter. The networks that helped shape this book are geographically oriented around the UK, Europe, USA, and Japan, leading to a stronger focus on these areas. The exhibition was a concrete starting point, a collection of art-objects that provoked further discussion, and it is intended that this collection of essays, by artists, curators, and art historians, calls for continued discussion, to extend beyond its geography, invite others into the conversation, and to draw attention to other 'radiant objects'.

What is meant by a 'radiant object'? The word 'radiant' denotes brightness and warmth and can call to mind images of rays shining from a source. It is often used to describe someone or something beautiful, or powerful. This book will look at how art has the power to turn our attention to, and help us reflect for longer on, the lived experiences of nuclear trauma. As such, these essays about encounters with objects, particularly art-objects, aim to help us perceive and comprehend the slow violence of nuclear radiation.

In this sense, we can think of radiant objects – like the holey hat – as being endowed with a form of agency. They are things that can be touched, which have materiality, while also having the ability to reconfigure our relationship to the nuclear, to help us grasp the enormity and pervasiveness of nuclear culture, and to better understand our subject position within the nuclear era.

Object-oriented ontology – the philosophy of being that understands objects as vibrant agents and not entirely 'knowable' to humans – offers us a set of tools to make sense of radiant objects' agentic role. It allows us to see how 'radiant objects' emit power and can enlighten those they encounter about lived experiences of the nuclear age, even as they elude complete understanding.

This approach also helps us to get to grips with the sheer magnitude and incomprehensibility of radiation. Timothy Morton uses the term 'hyperobject' to describe an object – like a black hole, or the Internet – that can only be understood in terms of fragments of an unknowable whole. Radiation might be considered a 'hyperobject': impossible to grasp in its totality, and accessible only through fragments such as the holey hat, the lens from a pair of glasses, an eraser (examples that will be considered). This collection draws on these aspects of object-oriented ontology to help us understand radiation as a 'hyperobject' that, although invisible to the human eye, has agency and can be encountered through tangible 'things'.

Towards a Philosophy of 'radiant objects'

As Timothy Morton reminds us in *Hyperobjects: Philosophy and Ecology after the End of the World*: 'There is no Goldilocks position that's just right from which to view objects' (2013, p. 36). Morton's articulation of the 'hyperobject' provides a helpful reference point when reading the following essays. A 'hyperobject' is an 'agent' (Morton 2013, p. 29) and so it is easier to discuss what hyperobjects *do*, or *how* they are, rather than what they are. Radiation, for example, is perceptible through its effects on things. Understanding that we live within hyperobjects, affecting and being affected by them, can help us approach issues such as radiated landscapes and environmental crisis with the knowledge that we are not 'in control', and there is no 'right view' from which to grasp them: they are both indeterminate and unknowable, while also being viscerally real. But if there is no 'right' view through which to understand a hyperobject, such as radiation, how can we begin to theorise, to think about and respond to it?

For Jorella Andrews, writing about the agency of objects in *Visual Cultures as Objects and Affects*, the concept of the 'shadow' can help us navigate the lack of visibility and clarity we have regarding an object (which is particularly relevant to Morton's 'hyperobjects'). Andrews discusses how a shadow is a visible testament to something concrete but itself is intangible: the 'doubled characteristic of the shadow, its connection with both the ungraspable and the concrete, suggests that incalculability and material reality are best not thought of as opposed …' (2013, p. 46). To accept indeterminability and concrete materiality as consonant allows for a more complex and nuanced interpretation of objects and affects. Crucially, holding this in our mind can, following Morton, help us focus on what objects *do*, rather than what they *are*.

It can be challenging to think of indeterminability and concrete materiality as compatible. Living in modernity, we are used to quantifying time, things and experiences into determinable 'values'. We want to measure hyperobjects, to find their limits, to be 'in control'. Systems of control rely on the articulation of 'values', equating objects and events that are deemed similar in function or meaning.

Although 'value' is a dynamic concept, shaped by social relations, in capitalist modernity it is often expressed as an 'exchange value'. Here, a thing's worth is expressed in terms of its value when traded for something else: very often, money. Exchange value permeates human experience and meaning-making, mediating our relationships. Philosopher Jean-Luc Nancy links the prevalence of exchange value to global culture in the years following the tsunami and nuclear disaster in Fukushima. In his 2015 book *After Fukushima: The Equivalence of Catastrophes*, Nancy asserts the idea of 'non-equivalence'. In simple terms, this means being able to perceive phenomena without ascribing values that bring them into the systems of 'general equivalence': a Marxist term that points to processes of totalising reality so that relationships between people, labour, and objects are expressed through money.

Nancy infers that trying to totalise disasters such as Fukushima through systems of quantification leads to the illusion that humans can be 'in charge' of the world (2015, p. 6, 7). He sees general equivalence as catastrophic. Nancy discusses how nuclear weapons create a 'balance of terror' that destroys the links between the strong and the less strong in society, and in doing so, closes the possibility of even being able to imagine a strategy to confront the force of nuclear catastrophe (2015, p. 22). The social atomisation caused by general equivalence, combined with the shift in the scales of power that nuclear technologies have brought, further amplifies inequalities and social disconnections between people and their environments. In this cultural context, rooted in fear and precarity, desires for control are perpetuated and indeterminacy can seem like a threat.

Following this, to acknowledge the indeterminate yet concrete quality of nuclear radiation contamination may offer little comfort or hope. However, accepting indeterminacy is crucial to sustaining human connection and responsiveness, even as we may try to evoke a concrete sense of an experience through writing or art. In *The Writing of the Disaster*, literary theorist Maurice Blanchot writes about disaster as 'what escapes the very possibility of existence – it is the limit of writing... the disaster de-scribes ... it is beyond the pale of writing' (Blanchot 1995, p. 7). Because disasters are of such magnitude, an experience of disaster cannot be communicated. For Blanchot, writing is passive and representative and can never fully communicate the sense of an experience. Writing gives meaning, but at the same time illuminates an absence of meaning. And yet it is important.

He quotes Nietzsche, who depicts words as having 'grazed a problem' (1995, p. 106). This simple analogy indicates that whilst writing cannot articulate, let alone solve a problem, it engenders sensitivity. In other words, the disaster is a kind of numbness to which writing alludes but cannot capture. It is important to acknowledge that the disaster exceeds our senses because to assimilate a disaster into reality is to reduce it and quantify it. In other words, to translate it into something that has 'general equivalence' is to orient thinking towards insensitivity and, ultimately, on the level of individuals' experiences, dehumanisation.

Because artists often approach concrete events with a sense of their incommensurability, they can help audiences develop a resilient comfort with the apparent paradox of indeterminability/concrete materiality. Often, this is experiential: it happens to viewers via visual language, rather than written text. However, art writing and art historical texts bolster and extrapolate these ideas, providing philosophical perspectives alongside some 'thing' to 'think with'. Art may not solve our problems quickly (or at all), but it can help us retain a sensitivity that can facilitate connection and care.

Gabrielle Decamous extensively examines the aesthetics and ethics of atomic-related art in her book *Invisible Colors: The Arts of the Atomic Age.* She reminds us of the ways in which critical engagement with the nuclear has been censored and self-censored. For Decamous, '[e]xamining the atomic age, together with the art created in response to it, allows us to see beyond, or find new resonance in, the tropes we have come to associate with it' (2019, p. 335). Recurrent motifs such as the use of the word 'clean' with relation to nuclear energy production, can blur our perceptions of nuclear materiality. Such images from the re-emergent nuclear dream can leave fuzzy sunspots in our understanding of nuclear cultural heritage. For Decamous, art is revealing (2019, p. 337) and carries with it an ethical imperative to retain and share the memories of lived experiences of the nuclear age.

The idea that artists are vital to understanding and responding to such experiences chimes with the words of Ele Carpenter, in *The Nuclear Culture Source Book*, where she writes about how '[a]rtists navigate the shifting timescales of environmental urgency in the present, planning and construction over generations, and deep time thinking over millennia' (2016, p. 15). The role of artists in thinking and responding to the indeterminable impacts of nuclear radiation can shape the way in which narratives of environmental crisis unfold and affect us.

In her book *Radioactive Ghosts*, literary theorist Gabriele Schwab writes about the 'nuclear unconscious' (2020). For Schwab, the impact of nuclear technologies is so huge that the human imagination is unable to grasp its significance. At the same time, she believes that the threat of nuclear trauma and the slow violence of the nuclear, including radiation contamination, cancers, and ultimately extinction of life on earth, lies heavily in our psyche, repeatedly pushed beneath our consciousness. This nuclear unconscious characterises what it means to be alive today. It forms a blurred, boundless backdrop to our everyday lives. Nevertheless, as objects of the nuclear imaginary come into focus, like the hat, there are sudden glimpses of the magnitude of nuclearism and its implications.

These glimpses can evoke a sense of powerlessness and of inevitability which can cause us to feel resigned, to turn our attention away from such objects. Although this is a natural human reaction, the rejection of nuclear consciousness is simultaneously a disavowal of both the lived experiences of physical nuclear trauma and its psychological impact. Choosing to turn away from this awareness can contribute to the erosion of the very humanness that shapes our cultures and relationships.

Our awareness of our nuclear subjectivity is reflected in, and shaped by, the cultures we live within. The following section begins to explore this cultural awareness, spoken of here as the emergence of 'nuclear cultures', particularly following the first atomic tests. The fear that Schwab writes about, combined with incomprehension of the concrete, but often gradual, impacts of nuclear production within social, medical and political spheres, has led to a sublimation of nuclear anxiety, one that shapes visual cultures. From *The Simpsons* to *Fallout*, imagery of the nuclear is commonplace, and even the focal point for popular films and series.

Nuclear Cultures

Although we may not live in the Springfield of *The Simpsons*, with its leaking nuclear power plant and three-eyed fish, we do still live within nuclear cultures. In her introduction to *The Nuclear Culture Source Book*, Carpenter describes this as 'the emergence of nuclear material cultures understood both as technological infrastructures and aesthetic practices' (2016, p. 9). Since the 1940s, such infrastructures have multiplied, and awareness of nuclear material culture has never been far from the surface of everyday life. This technology has informed and shaped visual culture; from the branding of crisps and lingerie to X-ray shoe-fittings and CND protest signs. Once attuned to this, it becomes more obvious how references to the nuclear suffuse through pop culture.

While narratives about radiation and its medical uses began in 1895 when Wilhelm Conrad Roentgen discovered X-rays, it was the development of nuclear weapons by the Manhattan Project from 1942 to 1946 that brought about one of the most significant shifts in human and geological history. For this reason, many contemporary cultural narratives have returned to the moment at which the first nuclear bomb was detonated, on July 16, 1945 in Los Alamos, New Mexico. This nuclear test – Trinity – has become symbolic of a new set of existential and moral questions, explored by many different artists, filmmakers, and writers. Here I would like to briefly draw attention to two different examples of how such pop-culture narratives facilitate critical attention towards nuclear history: the film *Oppenheimer* and the series *Twin Peaks: The Return.*

Christopher Nolan's 2023 film *Oppenheimer* tells the story of J. Robert Oppenheimer, the director of the Manhattan Project's Los Alamos Laboratory, which orchestrated the Trinity tests. The official Universal Pictures trailer begins with the words: 'We imagine a future, and our imaginings horrify us'. Referring to the atom bomb: 'They won't fear it until they understand it. And they won't understand it until they've used it. Theory will only take you so far …' (*Oppenheimer* trailer 2023). The rationale that shaped the invention of the atom bomb, and its moral intent, is examined in the film with equal measures of malaise and tenderness. In the film, the birth of nuclear culture, at this scale, is justified through the idea that theory and reality are opposing, separate, and to witness nuclear catastrophe correlates political conflict into a reciprocal bond, all sides united in fear.

The film tells how the production of the bomb was initially justified as the only way to stem the power of Nazi Germany, who surrendered before the bombs were completed. Instead, the bombs were used in the 'Pacific War'. On August 6 and 9, 1945, the USA bombed Hiroshima and Nagasaki, leading to the surrender of the Imperial Japanese forces on September 2, 1945. The loss of human life in Hiroshima and Nagasaki was legitimised as 'for the greater good'. This logic often lies beneath experiences of nuclear trauma, for example the role of soldiers in nuclear testing programmes. But in the film, the idea that rational deterrence is only impactful once humans have observed the devastating impacts of the bomb (and that this can lead to 'a greater good') begins to come apart under the viewers' gaze.

Another representation, and perhaps a more direct indictment, of the Manhattan Project can be found in David Lynch's depiction of the Trinity test in the cult series *Twin Peaks: The Return,* which pinpoints the detonation of the atom bomb as the origin of darkness and evil on the earth. Described by co-writer Mark Frost as the 'origin story' for the series as a whole (Reed, 2017), this visualisation of the detonation invites us to imagine being at the centre of the explosion, engulfed in fire, heat and rubble, to observe a point at which energy congeals into a golden orb, which later, held in the hands of a woman, reveals the face of the show's focal character Laura Palmer. Here, the distance and abstraction of the bomb is converted into a visceral and personal narrative, communicated through a hand-held object. Lynch and Frost famously foreground sense and feeling above realism within their work – for this reason, the depiction of the atom bomb in *Twin Peaks: The Return* is particularly evocative in how it communicates a sense of the world-changing, time-warping, and object-forming nature of the event.

I draw attention to these recent examples of nuclear visual culture because they invite viewers to contemplate the lived experiences of nuclear trauma in different ways – through the retelling of history *and* the use of fiction and fantasy. At the same time, they both ask questions about the ethics and practice of nuclear technology at this moment in time: a point in history at which the demand for nuclear energy is increasing along with the risks of nuclear catastrophe. Both examples communicate a sense of disconnect between humans' technological power and our comprehension of the long-term impacts of this power. These concerns are voiced through stories of individuals and attention to familiar and fantastical objects, allowing for a gradual and embedded perception of the complexities of nuclear cultures.

A 2022 study led by Lili Xia, an atmospheric scientist at Rutgers University, flagged how the impact of future nuclear weapon detonations would lead to mass food shortages in most countries, estimating that a nuclear war between the United States and Russia would cause more than five billion deaths. Although images of a 'nuclear winter' are easy to find in visual cultures (IMDB suggests a list of 35 films), a 2023 study at the University of Cambridge led by Paul Ingram and involving 3,000 participants indicated that that just 3.2% of people in the UK (and 7.5% in the US) had heard about this concept through contemporary media or culture. Despite increased risks of nuclear conflict and speculative imagery in visual cultures, there is little awareness of its impacts. This is perhaps even more the case when it comes to the 'slow violence' (Nixon 2011) of the impacts of radioactive waste produced by nuclear energy production. The 'invisibility' of radiation can allow for the proliferation of the myth of nuclear energy as 'clean energy', distracting us from its long-term insidious toxicity. But research indicates further ecosystem degradation through nuclear production, even as it potentially reduces air pollution (Soto and Martinez-Cobas 2024).

Facing the magnitude of nuclear capability and risk, we articulate our collective vulnerability through the image of the Doomsday Clock. Created in 1947 by scientists (including Oppenheimer), it indicates how close we are to a global nuclear catastrophe. Here, a clock – a domestic object invented to unitise time to a scale that makes sense to human rhythms and needs – becomes an uncanny signifier of human impacts on deep time. The Doomsday Clock is currently set at 90 s to 'midnight'. Despite the creation of this warning metaphor in the 1940s, the 1950s brought with it the 'nuclear dream', the sense that nuclear technology can hold a peaceful tension, simultaneously providing abundant 'clean' energy. Kitsch imagery from the 1950s focused on nuclear energy production and pitched the atom as a 'friend', heralding better health, new products, and more electricity. The nuclear meltdowns and radiation leaks in subsequent decades, the 'maximum severity' meltdowns and radiation leaks in Chernobyl in 1986 and Fukushima in 2011, undermined images of nuclear energy as 'clean'.

However, as our use of artificial intelligence snowballs and we require more energy than ever before, nuclear energy is hailed as one significant source that meets this demand. As such, we are witnessing the renaissance of the nuclear dream. Not only do we look to nuclear for more energy, but also for employment opportunities and social stability. Around the world, politicians promise new nuclear power plants as pathways to limitless energy and social cohesion. In a quest to finding a technofix for climate change and to increase energy production, governments have been pumping money into nuclear fusion development. In addition, the last decade has seen the emergence of small modular reactors (SMRs) and microreactors. While these are still being researched and developed, energy production companies are looking to shape public opinion through cultural discourses, inviting artists and writers to engage in 'critical debate', carefully overseen by the producers of these technologies. One example of this was a conference, bringing together artists, scientists, and philosophers, at Castello di Rivoli Museo d'Arte

Contemporanea in 2023, which was supported and co-hosted by the nuclear energy company Newcleo. The same year, the Nuclear Culture Research Group shared an open letter to artists, architects, curators, and the SMR research industry to 'demand that the arts reject funding from nuclear lobbies and support more inclusive and realistic discussions about Small Modular Reactors which value the lessons learned from the past and from wider cultural and political perspectives' (2023). The glamourisation of SMRs through cultural projects such as the conference at Castello di Rivoli Museo d'Arte Contemporanea, downplays histories of nuclear disaster, and the issues of nuclear waste disposal. The open letter addresses the implications of such 'art washing' as a form of censorship of voices that truly critically engage with the entanglement of nuclear technology, humans, and non-humans.

In the face of growing concerns about climate change, ever-increasing demands for energy and social instability, governments are expressing steadfast commitment to nuclear 'programmes', which include nuclear weapons as well as power plants. At the time of writing, the threat of nuclear weapons is increasingly in the media spotlight, as Israel strikes Iran's nuclear facilities, to stem the development of a nuclear bomb. Around the world, nuclear deterrents continue to be developed and/or maintained, and power plants are being reactivated, even those on the sites of past meltdowns and radiation leaks. Nevertheless, the benefits of nuclear technology are always 'to come' – the 'green' impact just a few years away, its promise of peace forthcoming.

Recalling the perceived justification for the Manhattan Project, the idea that the 'imaginings that horrify us' can be realised and learned from, it seems that we have pressed on to a stranger, more entangled position, one in which there is no simple response. As anthropologist Anna Lowenhaupt Tsing writes in her book *The Mushroom at the End of the World – On the Possibility of Life in Capitalist Ruins*, we mostly 'have to work within our disorientation and distress to negotiate life in human-destroyed environments' (2015, p. 131). She describes this as 'feeling around in the duff' (2015, pp. 278, 279). But how can we begin to think about, and respond to, this condition of perpetual uncertainty? And what role do objects and artworks have in this process of 'negotiation'?

'Nuclearity' and Nuclear Decoloniality

A significant term within studies of nuclear culture is 'nuclearity', used to convey a sense of the complexity of what it means for something to be categorised as 'nuclear'. In her book *Being Nuclear: Africans and the Global Uranium Trade* (2012), Gabrielle Hecht explains how the boundaries of what is 'nuclear' are contested and unstable (pp. 11, 12). Nations, technologies and materials can all be described as 'nuclear' but, as Hecht explains, '[t]here isn't one nuclear ontology; there are many' (2012, p. 14). To indicate the contested nature of how spaces, places and schemes might be 'nuclear', she uses the term 'nuclearity'. This term appears throughout this volume, spotlighting the varied qualities of nuclearity.

In mainstream media, discussions of nuclearity often centre on the impacts of nuclear weaponry and the latter stages of nuclear fuel cycle, including the damage of radiation and how to dispose of nuclear waste. This collection reflects this focus because many of the contributing artists are responding to public discourse, and to sites within their locale. And because this collection grew from an exhibition in the UK, and the geographic orientation of the book is the UK, Europe, the USA, and Japan. However, attention to nuclearity requires looking at the beginning of the nuclear cycle, the mining of uranium, which occurs mainly in the Global South and former Soviet nations, notably Kazakhstan.

At the time of writing, we are approaching the 80th anniversary of the testing of the first nuclear bomb and the bombing of Hiroshima and Nagasaki. The first nuclear bomb used uranium that had been mined from the Shinkolobwe mine in the Democratic Republic of Congo, at that time under Belgian colonial rule. The violence of colonial rule and dangers of working in the mines are addressed by Susan Williams in her book *Spies in the Congo: The Race for the Ore That Built the Atomic Bomb* (2016). Williams writes about the complex political context of uranium and the damage from the radiation on the local people in and near the mines. However, this, and many other histories of nuclearity, are often overlooked.

The effects of nuclear testing on Indigenous people are addressed by Greg Dvorak in his book *Coral and Concrete: Remembering Kwajalein Atoll Between Japan, America, and the Marshall Islands* (2018) where he recalls his childhood growing up in Kwajalein Atoll in the Marshall Islands in the 1970s. Between 1946 and 1958, the US undertook 67 nuclear tests in the Marshall Islands, primarily at Bikini and Enewetak Atolls. Dvorak writes about the impacts of colonialism on the local people, using coral and concrete as metaphors to understand the complex evolution of the area. For Dvorak, coral is a metaphor for the 'small, diverse, and marginalised' participants of history 'connected through the larger network of the atoll, and how these histories have sedimented into massive and powerful structures and continued to evolve' (2018, p. 23) in contrast to concrete as 'quick-drying colonialism' (2018, p. 25). For Dvorak, the 'concretising' effect of colonialism amalgamates and coopts these sedimentary stories, leading to a homogenised narrative (2018, p. 26). *Coral and Concrete* aims to provide access to alternative narratives that increase awareness of the experiences and culture of the Marshallese people.

The idea of coral as a metaphor for the 'small, diverse and marginalised' allows us to imagine something tangible, alive and evolving. Coral, part of the Cnidaria animal group, is alive, its secretions form a kind of skeleton of calcium carbonate. The rock-like quality we might associate with coral, its 'objectness' perhaps, is part of the coral's ecosystem, and as such, is 'vibrant matter' (to borrow a term from Jane Bennett's 2010 book, *Vibrant Matter:*

A Political Ecology of Things). Looking at how the marginalised might have agency, how objects have vibrancy, can open critical space for further discussions around nuclearity. Whilst the essays in this collection do not necessarily focus directly on colonial histories, they address the heterogeneity of nuclear cultures. Implicit in this is an understanding of the colonial power that has shaped nuclearity. Attention to the tactility and tangibility of the objects considered in this collection can help us understand the materiality of the histories they evidence, and the powers that have shaped these realities. Thinking of such objects – from pamphlets and erasers to buildings and even sounds – as 'radiant', as emitting energy, however small they may be, can be an entry point into unpacking nuclear decoloniality.

Overview of the Collection

The opening chapter of this collection is an essay by Atsuhide Ito that situates 'radiant objects' as references to absent bodies. Ito analyses objects including a leather belt worn by schoolboy Osamu Nanko on the day of the atomic explosion in Hiroshima, a pair of glasses worn by Shoji Yoshioka on the same day, and a piano found in the sea near the coastline of Fukushima. Through discussion of 'haunting' in Jacques Derrida's writing, connecting this to object-oriented ontology, Ito contemplates how objects can emanate a sense of presence and absence. He likens these objects to photographs in which a key reference has been removed. Ito analyses this complex moment of encounter to forward deeper engagement with such articles.

Building on ideas of encounter, and through the photograph itself as an object, artist Yoi Kawakubo reflects on the process of creating a series of photographic prints within the Fukushima exclusion zone. Focusing on the creation of one particular work: *If the Radiance of a Thousand Suns Were to Burst at Once into the Sky*, an enlarged print formed from photographic film that had been buried in irradiated soil in the restricted zone, Kawakubo's account weaves together his personal narrative, art historical context, and critical reflections on the imperceptibility of nuclear contamination. This essay explores how we might begin to engage with the materiality of radiation, despite its intangibility.

Developing this idea of making the intangible perceptible, the following chapter explores the creation of 'noise-prints', a concept created by artist Daniel Beck as part of his research into the British Film Institute (BFI) storage facility at Gaydon in Warwickshire, a former nuclear storage facility. This Cold War era depository site, often overlooked in histories of nuclear culture, becomes central to Beck's enquiry into how film media, archive, and sound practice can become carriers of nuclear culture. Beck describes noise-prints as 'wayfinding instruments' that help us navigate through nuclear histories. He asks us to consider how the act of listening can be a starting point for new narratives of the nuclear and its impacts.

The theme of materiality is addressed by artists Bridget Kennedy, Robin Tarbet, and Helen Grove-White, with relation to the decommissioned Wylfa Nuclear Power Station. Written as a conversation, these artists also draw upon the writings of Gabriella Hecht, and the term 'nuclearity', to consider narratives of erasure surrounding nuclear culture. The discussion centres around a small, rainbow-striped rubber featuring the image of the power station. This eraser, once a souvenir in the gift shop, becomes a symbol of the ways in which nuclearity can be reduced, branded, and packaged into something malleable and kitsch, an object that can be absorbed into the realm of the domestic. Kennedy, Tarbet, and Grove-White take this object as a starting point to consider nuclear decommissioning projects, the future of nuclear, and the role of artists in responding to the political and ethical dilemmas of nuclear industrial projects.

In response to these political and ethical dilemmas, alternative approaches to energy production have proliferated. Interdisciplinary researcher, artist and activist Marc Herbst turns our attention to creative responses to some of the issues addressed in the previous chapters. Turning to a small, printed book pamphlet: *Einfälle Statt Abfälle* (*Inventions Instead of Waste*), created by Christian Kuhtz in 1983 and promoting DIY energy projects, Herbst examines the relationship between individuals and communities, states and social movements, in the wake of nuclear energy. Although small, this publication can be interpreted as a kind of alternative monument to the legacies of German nuclear history. Herbst draws upon the pamphlets to reflect on the development of the German anti-nuclear movement and the boundaries between state power and resistance.

From Germany to Albania, the next essay by Dimitra Gkitsa reflects on the nuclear bunker as a 'radiant object'. Throughout the 1970s and 1980s, there was a proliferation of bunkers in Albania, overseen by the country's communist party. When the regime collapsed, these bunkers fell into neglect and many were vandalised, forming a visual reminder of traumas experienced under the rule of the Party of Labour of Albania (PLA). Gkitsa points to examples of 'dark tourism' in such sites: two of the largest nuclear bunkers, located in Albania's capital, Tirana now contain history museums and gallery spaces Bunk'Art 1 and Bunk'Art 2. Gkitsa critically engages in the ways in which communities and artists repurpose nuclear cultural heritage sites such as these to reshape histories and cultural narratives.

Continuing with the context of the Cold War, the subsequent chapter critically engages with the site of Britain's nuclear weapons site, Faslane in Scotland. Home to four nuclear submarines, HMS Resolution, HMS Repulse, HMS Renown, and HMS Revenge, the surrounding water contains radioactive isotopes. Artist Michael Sanders critically responds to the complexity of this site through the creation of 'contaminated tartans' that make the nuclear history of Faslane visible through encoded textile design. This chapter, written in collaboration between Sanders and fashion historian Jonathan Faiers, spotlights symbolism within the history and future of tartan. Here tartan becomes a tool in both camouflaging and exposing nuclear culture.

In the next essay, the invisibility/visibility of nuclearity is explored in a different historical context. Artist, activist, curator and mental health counsellor Tim Andreae has spent over a decade making artworks and exhibitions in response to the Idaho National Laboratory (INL) and the conundrums presented by the existence of radioactive waste and its threat to life on the Snake River Plain. Andreae identifies the INL as a place whose strivings and contrivings encapsulate the sickness of our age. In this essay, he takes up a single thread of the nuclear endeavour – plutonium-contaminated waste, or 'transuranic waste' – and explores its mismanagement at the INL through historical, psychological, and mythic lenses. He invites us to consider the toxic byproducts of our nuclear strivings as expressions of our collective shadow. Here, the 'shadow' cast by the hyperobject of plutonium can also be understood in Jungian terms, as part of our collective neurosis.

The penultimate essay in the collection again forwards art as catalyst in understanding and healing, adding to this the importance of connection with the more-than-human. Returning to Fukushima, Jason Waite approaches the exhibition Don't Follow the Wind as a cultural object. This exhibition, based inside the currently inaccessible exclusion zone in Fukushima, is accessed primarily by non-humans. The focal image for this chapter is a photograph of a Palm-faced civet encountering Trevor Paglen's Trinity Cube. Waite draws on 5 years of research with scientists, local foresters, and hunters in the region to contest the logic of 'nuclear re-wilding' and to pay attention to the more-than-human life that inhabits this degraded environment.

Concluding the volume, PhD researcher Min-Kyoo Kim focuses on the 'immateriality' of digital archives and the Internet, reflecting on the opportunities and challenges for the preservation of 'radiant objects' amidst the proliferation of virtual archives. Kim connects Derrida's Nuclear Criticism with Annie Jacobsen's Nuclear War, to imagine how the electromagnetic pulse (EMP) of a nuclear attack could potentially sever our relationship to the digital archive. Extending Derrida's analysis of survivance to include visual media beyond the literary, Kim questions how and why we might preserve the archive. This concluding reflection spotlights the materiality/immateriality of objects within the collection, to further comprehend the potential precarity of radiant objects that anchor our relationships to the past and to each other in the enduring Atomic Age.

Encounters, between humans and more-than-humans (animals, objects, hyperobjects), and even between objects themselves, underpin this collection of writings. This volume aims to give insight into such meetings and encourage readers to be receptive to experiencing the unknown and unknowable. One way of facilitating encounters, particularly those that make us more aware of issues that feel overwhelming and irresolvable, is through engagement with art. One of the aims of this collection is to allow the reader to contemplate the difficult issues that we collectively face for a little longer, and with attention to the quieter voices and hushed narratives of the nuclear age.

References

Andrews, Jorella & O'Sullivan, Simon. 2013. Visual Cultures as Objects and Affects. Berlin: Sternberg Press.

Bennett, Jane. 2010. *Vibrant Matter: A Political Ecology of Things*. Durham, NC: Duke University Press.

Blanchot, Maurice. 1995. *The writing of the disaster: L'écriture du désastre*. Translated by A. Smock. Lincoln: University of Nebraska Press.

Carpenter, Ele. [Ed]. 2016. *The Nuclear Culture Source Book*. London: Black Dog Press

Decamous, Gabrielle. 2019. *Invisible Colors: The Arts of the Atomic Age*. Cambridge, MA: MIT Press.

Dvorak, Greg. 2018. *Coral and Concrete: Remembering Kwajalein Atoll Between Japan, America, and the Marshall Islands*. Honolulu: University of Hawai'i Press.

Hecht, Gabrielle. 2012. *Being Nuclear – Africans and the Global Uranium Trade*. Cambridge, MA: MIT Press.

Ingram, Paul. 2023. Cited in University of Cambridge Research. Available at https://www.cam.ac.uk/research/news/public-awareness-of-nuclear-winter-too-low-given-current-risks-argues-expert [Accessed 6/25].

Morton, Timothy. 2013. *Hyperobjects: Philosophy and Ecology after the End of the World*. Minneapolis, MN: University of Minnesota Press

Nancy, Jean-Luc. 2015. *After Fukushima: The Equivalence of Catastrophes*. Translated by C. Mandell. New York: Fordham University Press.

Nuclear Culture Research Group. 2023. *Open letter on Art Washing Small Modular Reactors*. Available at: https://docs.google.com/forms/d/14ei1Q7PfrMiHOOUNbCRWaY4oCZSj1EDYvlmMZW_94mc/viewform?ts=65841 8a3&edit_requested=true [Accessed 7/25].

Nixon, Rob. 2011. *Slow Violence and the Environmentalism of the Poor*. Cambridge, MA: Harvard University Press.

Nobody Told Us Anything. 2014. [film]. Directed by Charles Stewart.

Oppenheimer. 2023 [film]. Directed by Christopher Nolan. USA. Universal Pictures.

Oppenheimer trailer. 2023. Available at: https://www.youtube.com/watch?v=keydELfWuMQ [Accessed 6/25].

Radiant Objects: encounters in the nuclear age. 2023. Newark Works. Fringe Arts Bath. Bath. Info available at: https://www.fringeartsbath.co.uk/radiant [Accessed 07/25].

Reed, Ryan. 2017. 'The last word on "Twin Peaks" by David Lynch's co-creator Mark Frost', Salon. 7/11/2017. Available at: https://www.salon.com/2017/11/07/the-last-word-on-twin-peaks-by-david-lynchs-co-creator-mark-frost/ [Accessed 06/25].

Schwab, Gabriele. 2020. *Radioactive Ghosts: Volume 61*. Minneapolis, MN: University of Minnesota Press.

Soto, Gonzalo H. & Martinez-Cobas, Xavier. 2024. Nuclear energy generation's impact on the CO2 emissions and ecological footprint among European Union countries, *Science of The Total Environment*, Volume 945, 1 October 2024, 173844.

Tsing, Anna Lowenhaupt. 2015. *The Mushroom at the End of the World – On the Possibility of Life in Capitalist Ruins*. Princeton: Princeton University Press.

Twin Peaks: The Return, Episode 8, Got a Light? [TV] Directed by David Lynch. Showtime. June 25, 2017.

Williams, Susan. 2016. *Spies in the Congo: The Race for the Ore That Built the Atomic Bomb*. London: C Hurst & Co Publishers Ltd.

Xia, Lili, Robock, Alan, Scherrer, Kim *et al*. 2022. Global food insecurity and famine from reduced crop, marine fishery and livestock production due to climate disruption from nuclear war soot injection. *Nature Food*, Volume 3, 586–596. Available at: https://www.nature.com/articles/s43016-022-00573-0 [Accessed 6/25].

1 Radiant Memory

Atsuhide Ito

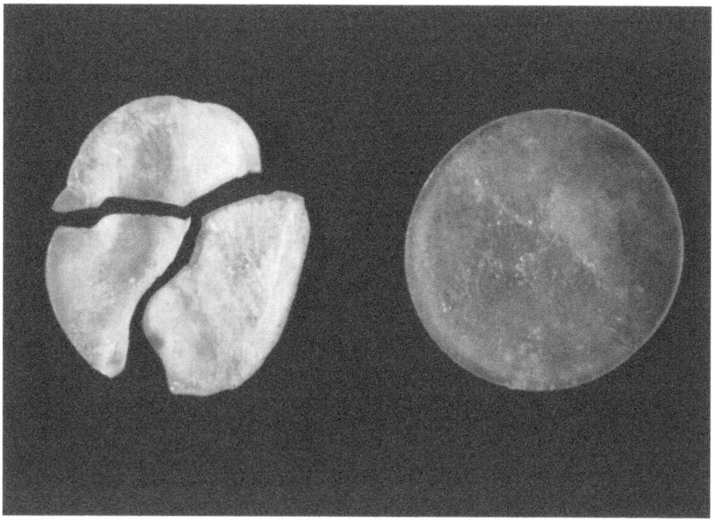

Figure 1.1 Shoji Yoshioka's glasses. Donor: Masako Arii. Collection of Hiroshima Peace Memorial Museum.

Radiant Objects

On August 6, 1945, a 31-year-old police officer, Shoji Yoshioka, travelled by train to collect some documents from the prefectural government office in Hiroshima. On his way back, he felt sick and decided to visit a hospital in Sakaimachi, in Hiroshima's city centre. I imagine he waited for a while until he was invited into a room. The doctor, perhaps a young man, asked him to lie down on a modest bed for an examination. While the doctor filled in a form at his desk, Yoshioka stared at the ceiling blankly. Suddenly, with a flash of light, his vision was violently struck and the lens of one of his round glasses

DOI: 10.4324/9781003622185-2

pressed deeply into his eye, driving it into his skull before breaking into three pieces. Simultaneously, the hospital went ablaze, and with it, the other lens of Yoshioka's glasses melted; minute bubbles formed on the surface, abruptly losing their translucency. Viewers of the objects, decades later, may be oblivious to the levels of radiation retained by the glasses (shown in Figure 1.1).

The eyes behind the glasses witnessed the atomic explosion that occurred at 8:15 a.m. in Hiroshima in 1945. Viewers of the objects, decades later, may be oblivious to the levels of radiation retained by the glasses. Here, the term 'radiant' in 'radiant object' does not refer to the object's irradiation. Instead, the radiant object becomes symbolic in the mind of the spectator; it is a projection of the viewers' understanding of the object as a legacy of the atomic bomb and the absent pair of eyes that once wore them. Yoshioka was missing after the explosion. After digging through the burned-down hospital grounds, his family was able to locate his glasses as a *memento mori.* In this sense, the patina of Yoshioka's glasses is both a tragic reminder of the missing body and a medium through which to mourn the missing person. I seek to develop this point further in the following example.

The Hiroshima Peace Memorial Museum archives torn up dresses, school uniforms, and a portion of rice-turned charcoal in a metallic lunch box, along with many personal items which used to belong to the victims who died from the explosion's impact and high level of radiation. Among these items lies Osamu Nanko's leather belt. Nanko was a 12-year-old junior high school student. On August 6, 1945, he was taking part in the task of demolishing a building. It was a sunny day; the sun was beating down on him. I imagine him to have been pushing a wheelbarrow filled with a pile of discarded building materials at the moment of the atomic explosion (*Hiroshima Peace Memorial Museum Collection Catalogue 2020*). His body suddenly rendered weightless, was blown away, only waking up under burning rubble. He was confused and disoriented but felt unbearable pain all over his body. His shirt and trousers were burned. With considerable difficulty, he got up and took off his shirt and trousers. Remembering his parents gave him the leather belt as a gift when he progressed onto junior high school, he held the belt in his hand firmly. He started walking slowly only wanting to go home, but all the buildings around him were destroyed or burning and he felt lost and disoriented, and he felt as if he was still burning. He frantically walked and walked as if to shake off the pain, but he became exhausted and decided to sit down, trembling on the side of a road. As it was getting dark, a man came to him and asked for his name and address. The man helped him walk and brought him home. His mother and brother opened the door and saw Osamu severely burned, wearing only underpants, holding his leather belt still firmly in his hand. He died a few hours later on the same night, in the company of his mother and brother. When visitors of the museum look at the belt decades later, though apparently the raggedness of it indicating its age, the signification of it makes the belt symbolically radiated. Unless the viewer is made aware that the belt was once worn by the boy whose traumatic experience is conveyed through the object and placed in a museum collection, the belt as a piece of leather would not immediately draw the viewer's attention. It is the thought of the absent body triggered by the silent belt that once enwrapped the body of the twelve-year-old. This structure of capturing absence is embedded in the radiant object.

Sigmund Freud (2005), in his short article 'Mourning and Melancholia', analyses the process in which a subject is affected by the loss of a loved one (object). The object can mean anyone who is significant to the subject. In his analysis, the loss of the object begins to affect the subject whose sense of loss increases and in turn empties out the subjectivity of the person. The loss of the loved one/object casts an internal shadow in the person whose attachment for the other consumes the subject and makes the person abandoned and lost. In Freud's own words,

> Der Schatten des Objekts fiel so auf das Ich, welches nun von einer besonderen Instanz wie ein Objekt, wie das verlassene Objekt, beurteilt werden konnte.
>
> (Freud 1917, p. 4)

> In this way, the shadow of the object fell upon the ego, which could now be condemned by a particular agency as an object, as the abandoned object.
>
> (Freud 2005, p. 209)

The shadow of the radiant object, such as the belt discussed above, consumes the viewing subject as if overshadowing them with the trauma of nuclear history. The belt signifies the lost object/Osamu Nanko, who can cast a shadow. In this manner, the symbolic power of the belt affects the viewing subject. The object here is both the outline of something lost, and, concurrently, is the physical object that triggers consummation of the viewing subject.

When discussing the radiant object, it is worth considering the psychoanalytical dimension to the term object. Jacques Lacan in his chapter, 'The Object and The Thing', remarks that 'in [psycho]analysis, the object is a point of imaginary fixation which gives satisfaction to a drive in any register whatever' (Lacan 1992, p. 113). In other words, the object is in one's imagination and drives the subject's actions and behaviours. In the same chapter, Lacan elucidates that this object of psychoanalysis is not the same as a physical object in a collection of objects (Lacan 1992, p. 113). Understandably, Lacan makes it clear that the object in psychoanalysis is not the same as a physical object. However, considering the psychoanalysis' relationship to collections of objects such as Freud's own collection of objects in his study, they cannot be simply excluded from an analysis. For instance, an object in an atomic museum, affects the viewer in the structure of mourning and melancholia. In *Four Fundamental Concepts of Psychoanalysis* (1994), Lacan deploys the expression 'object petit *a*', a psychologically unresolved element that drives the subject repeatedly to look for it. Lacan explains that such a repression produces an unattainable missing object, object petit *a*, for which the subject pursues and desires (Bourriaud 2022, p. 102). When applying the structure of pursuit in search of the object in collective memory and history, the radiant object is, I argue, a psychoanalytical object proper in the sense of object petit *a*.[1] To think of a radiant object with a psychic dimension helps evade the tendency to see an object as a physical thing that is made static and stable. For Lauren Berlant, 'objectness is only a semblance, a seeming, a projection effect of interest in a thing we are trying to stabilise' (Berlant cited in Yusoff 2022, p. 13). When a thing is placed in a museum, despite its potency of evoking and provoking experiences and symbolic suggestions, it is made to fall under the gaze of the viewer for inspection and contemplation; in this process the thing becomes fixed, tamed and turned into an object of contemplation. In turn, things are captured and imprisoned behind glass in a museum display, thus becoming objects. Yet, the notion of the radiant object resists this stabilisation and fixation. Not only are the radiant objects irreducible and 'live' in their material time, but also thinking through the roles of the radiant object in the process of mourning helps us see the potency of the radiant object and its ability to connect the viewer to the phantom presence of the absent bodies. Similar to John Harries' (2017) notion of tactile memory, radiant objects evoke the memory of someone in a distant past, in a distant location.

Becoming Radiant

After the nuclear meltdown at the Fukushima Nuclear Power plant in 2011, a piano was found in the sea near the coastline of Fukushima. Having heard the news of the salvaged piano, the composer and musician Ryuichi Sakamoto, at the time living in New York, travelled to Japan. He played the now out-of-tune piano at a modest concert held in a school gymnasium in Miyagi. By the time Sakamoto played the piano, he had been suffering from terminal cancer for some time. In a documentary film entitled *Coda* (2017), Sakamoto analogously compares the out-of-tune piano with his deteriorating health. Despite his illness making his body 'out of tune', it also lets him realise how one's health in modern society requires rigorous management and maintenance of the body. It is almost as if Sakamoto recognises the death drive, Thanatos – the force that makes an organism return to the inorganic state, a death – operating in the piano and in him. After his passing, the piano came to embody not only the collective memory of the nuclear accident but also retains the memory of the composer's absent body, and the discordant sounds that once emerged from it. The haunting evokes signification, but it is through the viewer's awareness of the absent body missing from the object. This structure of signification operates in a similar way to that of a portrait photograph, in terms of the indexicality of a photograph.

In *Camera Lucida,* Roland Barthes (2000) discusses the photograph of his mother, aged five, taken in a winter garden in 1898. She is standing next to her brother and appears angelic. While talking about the image of his mother, Barthes does not offer the reader an opportunity to see the photograph. This coerces the reader to realise the indexicality of photography, which is structured through a missing referent. In this particular case, the focus is on his mother, who is the protagonist of the discussion yet remains absent – 'avisual', to borrow Akira Mizuta Lippit's term – and is simultaneously presented as an idea (Lippit 2005, p. 84, 92, 102). In other words, there is a contextual knowledge in the mind of the viewer that the absence, or lack, in the image, is actually the content of the image. A photograph leaves a trace of the person who is missing and yet this absence is structurally central to the photograph, the object petit *a.* Photography is a medium of mourning for the person who was caught in the moment that never returns. According to Barthes, 'the photograph is literally an emanation of the referent. From a real body, which was there, proceed radiations which ultimately touch me' (Barthes 2000, p. 80).

Carol Mavor propounds that 'a photograph is a trace of the death of the moment held forevermore' (Mavor 2012, p. 68). The statement epitomises the description of viewing the objects in atomic museums. Such objects are 'anasemic', to borrow Nicholas Abraham's (2012) term, enabling a 'search for the meaning of phenomena in something that is by nature inaccessible to direct apprehension' (Abraham cited in Mavor 2012, p. 19). Semiotics as a study of signs is insufficient to extract the degree of nuclear violence conveyed in those objects, and thus those objects are anasemic, and radiant in the sense of emanating far beyond the moment of the trauma. Radiant objects gesture back to the moment of nuclear trauma and injury and yet radiate powerfully to posterity for the generations later. The absence of a referent is not the same as lacking references, rather, it is the recognition of a missing referent that constitutes the object. This structure of haunting, or avisuality, characterised by the missing body as a referent, produces the radiant object – like an echo – or ionising radiation resonating into the distant future. Like a portrait photograph in which the photographed person is physically missing but represented as an idea and an image, radiant objects are made radiant through the missing body which once was intimately in contact with the object.

Byun-Chul Han says, 'photography is an ectoplasm' (2022, p. 30, 96).[2] Han's use of the term ectoplasm, an emanation of energy, is fitting when recalling the image of the bowl of radium bromide taken in 1922. 'Ect', as opposed to 'endo', suggests externality instead of internality. Thus, ectoplasm suggests becoming externally visible of plasma, materiality of both organic and inorganic forms. A conventional analogue photograph inscribes light and shadow onto film, which is then printed on photographic paper. It typically captures an object or figure illuminated by sunlight or another external light source. However, the photograph of a bowl containing radium bromide is a peculiar photograph. As the title of the photograph explains '*photographiée*

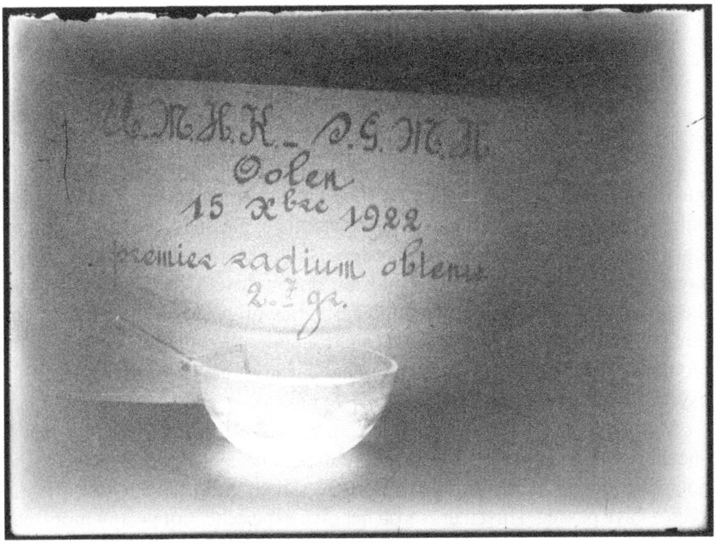

Figure 1.2 A bowl containing radium bromide photographed in the dark, 1922. Musée Curie (Coll. ACJC).

dans l'obscurité' (photograph taken in the dark), this image is of light itself, a form of radioactivity, emanating like a halo (see Figure 1.2). While the bowl provides a figurative form to the formless light, the key figure in this photograph is the light emanating from the radium bromide, which makes the image possible. Without external light, the source of light itself is the protagonist of the image. It presents itself by lighting itself. In so doing, it (radioactivity) lets itself be photographed, an auto-portrait of radium. In turn, the radium allows shadows to manifest, producing what Lippit calls 'skiagraphy, a shadow photography' or 'the shadow of photography' (Lippit 2005, p. 95).

Radiant Memory

The relationship between atomic explosions and photography is described in a spectacular manner in the writing of Paul Virilio who states, 'the bombs that dropped on Hiroshima and Nagasaki were *light-weapons*' (Virilio 1989, p. 101). Following Virilio, Lippit suggests:

> the atomic blasts and blackened skies can be thought of as massive cameras, then the victims of this dark atomic room can be seen as photographic effects. Seared organic and nonorganic matter left dark stains, opaque artifacts of once vital bodies, on the pavements and other surfaces of this

grotesque theatre. The "shadows" as they were called, are actually photograms, images formed by the direct exposure of the objects on photographic surfaces.

(Lippit 2005, p. 94)

Analogue photography relies on an external light source, typically the sun's rays, to locate and outline the figure or landscape to inscribe the image onto a film. However, in the case of the atomic explosions, the bombs' explosions are perceived as the surrogate sun. By the same violent gesture, Lippit suggests that the moment of explosion and its excessive light obliterated the sensibility and sensitivity to the aesthetics of shadows in Japanese culture. Like the moment of pressing a shutter button, with a thousand suns, the moment of explosion erased the physical shadows and instead produced the psychological shadows of objects, radiant objects.

Marianne Hirsh delineates 'postmemory' as a form of memory experienced by the generation after traumatic experiences have occurred – despite not being their own memories or experiences, their impact is felt long after the events have passed. Hirsch draws on her own experiences of postmemory, particularly in relation to the holocaust, as she bears the memories of her parents' traumatic experiences. Consequently, memory is transmitted and experienced as her own (Hirsh 2012, p. 5).[3] It is worth noting Hirsh's choice of word 'experiences' instead of a traumatic 'event'. It is not the moment of explosion but the durational legacy as slow violence (Nixon 2011). For Hirsh, postmemory is 'a *structure* of inter- and transgenerational return of traumatic recall' (Hirsh 2012, p. 6). By viewing a radiant object, the spectator is placed and allowed to weave postmemory of which they did not experience. Through this experience, the spectator may develop an insight into the devastating depth and scale of violence caused by atomic weapons. Through this process, empathy and the psychological transference of the suffering of the victim conflate, enabling the viewer to sensorially be drawn into a memory. However, in this process, different from the ways in which Hirsh outlines the structure of postmemory, the repressed point of trauma conveyed in the radiant object resembles the long-lasting and imperceptible impact of ionising radiation. Radiant objects evoke something of a 'radiant memory'. I would delineate radiant memory as a contaminated form of memory beyond the scale of the Anthropocenic time frame, and radiant objects as a catalyst for evoking such memory. The radiance of radiant memory encompasses colonial histories, uranium mining and nuclear weapons testing in remote former colonial territories of nuclear nation-states (Hashimoto 2003; Hecht 2012; Decamous 2018). Although the visitor of the museums does not necessarily connect those dots and concealed lines that link those dots immediately, they touch on this repressed and hidden violence that caused and still is causing trauma into the future beyond the Anthropogenic time scale. Like a photograph, radiant objects such as the Osamu Nanko's belt, the salvaged piano, and the melted lens allow the viewer to structure the relationship between them and those who once wore the belt, played the keys of the piano and wore the glasses. Radiant objects resemble photographs and allow the repressed psychological object to appear, letting them be internalised in the subject of the viewer. The process of finding imbrications with others' memory allows an insight into others' contemporary situations and struggles. Hirsh insists that postmemory contributes to constructive sharing of trauma through generations and across peoples (Hirsh 2012, p. 21). Radiant objects in this sense produces radiant memory that resonates epochally and inter-generationally and demands gestures of return, similar to haunting. As Jacques Derrida (1994) writes, haunting is a way to work for justice, instead of a passive acceptance of being implicated in the violence. The spectator's resonance with the radiant objects evokes a sense of reverie, or they 'reverberate' (Hirsh 2012, pp. 211–212). Furthermore, radiant objects suggest the presence of the missing referent for the viewer and activate the lines of memory from past to the future into which overshadowed atomic history extends. The radiant objects entangle the viewer into the burden of nuclear history. In other words, haunting produces relations that entangle the viewer into the unfinished nuclear history reminding the viewer that the u-chronia and u-topia of nuclear futurities are dreams shadowed by trauma. The radiant object is a medium to let us acknowledge the presence of the traumatised other in ourselves, that is a victim and aggressor in one.

Conclusion

'Photographs are a means to return to a moment or place in the past that troubles their own memories' (Kuhn and McAllister 2006, p. 2). In a similar manner, objects collected in atomic museums are mediums of return. Those objects provide a passage for a return to others' experiences. The signification of radiant objects relies on the memory of the missing referent, the body of the person who once wore and attached physically to the objects. The apparent absence of the body produces a lack evoking a passage of return to the traumatic experiences in the past, and yet, this process allows the viewer of the radiant objects to be entangled in the symbolic radiance emanating from the objects. Such radiant memory casts a long shadow over generations and by the same gesture eclipses the viewers of the radiant objects into the nuclear history of others as their own. And yet, the shadow, like the unconscious, provides a refuge and reconciliation from the atomic light of radiant memory.

Notes

1 Earlier in Lacan's seminars between 1959 and 1960, published in English under the title *The Ethics of Psychoanalysis*, there is no mention of object petit *a*, but Lacan discusses *Das Ding* (the Thing) in a sustained manner. It appears that Lacan's attempt to articulate his thought on The Thing led later to name unresolved

psychological thing, the object petit *a* in *Four Fundamental Concepts of Psychoanalysis* (1994, published in French in 1973).

2 In the first quotation, Han refers to photography as an ectoplasm and suggests it is an emanation of past reality. In the second quotation, Han is referring to his juke box as a specifically meaningful item in his life but here the second quotation is made relevant to consider the point of emanation.

3 'Postmemory: describes the relationship that the 'generation after' bears to the personal, collective, and cultural trauma of those who came before—to experiences they 'remember' only by means of the stories, images, and behaviours among which they grew up. But those experiences were transmitted to them so deeply and affectively as to seem to constitute memories in their own right' (Hirsh 2012, p. 5).

References

Barthes, Roland. 2000 [1981]. *Camera Lucida*. London: Vintage.

Bourriaud, Nicolas. 2022. *Inclusions: Aesthetics of Capitalocene*. London: Sternberg Press.

Decamous, Gabrielle. 2018. *Invisible Colors: The Arts of the Atomic Age*. Cambridge, MA: MIT Press.

Derrida, Jacques. 1994. *Specters of Marx: The Work of Mourning and the New International*. Trans. P. Magnus. London: Routledge.

Freud, Sigmund. 1917. 'Trauer und Melancholie', in *Studienausgabe, Bd. III: Psychologie des Unbewussten*. Frankfurt/M.: Fischer 19757 (1917), pp. 194–212.

Freud, Sigmund. 2005. 'Mourning and Melancholia', in Freud's *On Murder, Mourning and Melancholia*. London: Penguin.

Han, Byung-Chul. 2022. *Non-Things: Upheaval in the Lifeworld*. Trans. D. Steuer. Cambridge: Polity Press.

Harries, John. 2017. 'A stone that feels right in the hand: Tactile memory, the abduction of agency and presence of the past', *Journal of Material Culture*. Vol. 22, Issue 1. London: Sage. Available at: https://doi.org/10.1177/1359183516679187 [Accessed 3/25].

Hashimoto, Isao. *1945–1998*, 2003. Video 14:24 mins. Available at: https://sites.saic.edu/earthlyobservatory/artists/isao-hashimoto/ [Accessed 2/25].

Hecht, Gabrielle. 2012. *Being Nuclear: Africans and the Global Uranium T*rade. Cambridge, MA: MIT Press.

Hiroshima Peace Memorial Museum. 2020. *Hiroshima Peace Memorial Museum Collection Catalogue*. Hiroshima: Hiroshima Peace Culture Foundation.

Hirsh, Marianne. 2012. *The Generation of Postmemory: Writing and Visual Culture After the Holocaust*. New York: Columbia Univ. Press.

Kuhn, Annette and McAllister, Kirsten Emiko. 2006. 'Locating Memory: Photographic Acts—Introduction', in Kuhn and McAllister (eds.), *Locating Memory: Photographic Acts*. Oxford: Berghahn.

Lacan, Jacques. 1992. *The Ethics of Psychoanalysis 1959-1960: The Seminars of Jacques Lacan Book VII*. Trans. D. Porter. London: Routledge.

Lacan, Jacques. 1994. *The Four Fundamental Concepts of Psychoanalysis*. London: Penguin.

Lippit, Akira Mizuta. 2005. *Atomic Light (Shadow Optics)*. Minneapolis: University of Minnesota Press.

Mavor, Carol. 2012. *Black and Blue: The Bruising Passion of Camera Lucida, La Jetée, Sans soleil, and Hiroshima Mon Amour*. London: Duke University Press.

Nixon, Rob. 2011. *Slow Violence and The Environmentalism of the Poor*. Cambridge, MA: Harvard University Press.

Virilio, Paul. 1989. *War and Cinema: The Logistics of Perception*. London: Verso.

Yusoff, Kathryn. 2022. 'Epochal Aesthetics: Affectual Infrastructures of the Anthropocene', in Nick Axel, Daniel A. Barber, Nikolaus Hirsh, Anton Vidokle (eds.), *Accumulation: The Art, Architecture, and Media of Climate Change*. e-Flux Architecture, Minneapolis: University of Minnesota Press.

2 Photography and Materiality

Accounts from the No-Return Zone of Fukushima

Yoi Kawakubo

Figure 2.1 Installation view of the work at the exhibition *Vision of Contemporary Art 2015*, The Ueno Royal Museum. Image © Yoi Kawakubo.

If the Radiance of a Thousand Suns Were to Burst at Once into the Sky

In the historic city of Kurashiki, in the western region of Japan, nestled among rows of weeping willows, traditional low-rise buildings, and the serene historic canals once traversed by merchants and now mostly by tourists, stands

DOI: 10.4324/9781003622185-3

the Ohara Museum of Art. The museum is a large, square structure, constructed from granite stone blocks and partially concealed behind ivy-covered stone walls. Its solemn and heavy façade resembles that of a small Greek temple. Its triangular pediment is supported by four substantial columns, combining elements of Ionic and Etruscan styles. Completed in 1930, the museum houses Japan's oldest collection of Western art, including works by Picasso, Monet, and one of the country's only two El Grecos. The object of focus lies in one of the museum's more recent extensions. In a generous gallery space measuring approximately 15 × 10 m^2, with a moderately low ceiling, various works, mostly paintings, are hung sparsely along the walls. On one of these, a piece stands out due to its vertical composition, 3 m high and 2.5 m wide (Figure 2.1).

The artwork appears to float a few centimetres off the wall, composed of two Plexiglas panels, each about 5 mm thick, joined at the centre. At first glance, the strictly geometric rectangle resembles an abstract blotch of orange. At its core lies a vivid vermilion stain, radiating outwards into scarlet, then blending into a pale, almost greenish yellow near the edges. The Plexiglas reflects light and shadow across the space like a flawless mirror. A narrow white border, less than an inch wide, separates the coloured area from the Plexiglas edges. In the upper left corner, three small semi-oval notches disrupt the otherwise perfect geometry. Upon closer observation, finer details emerge: the colour field fills the entire visual frame, and the rectangular composition recedes into the background. The colour, which initially seemed smooth and uniform, reveals a granular and sandy texture, with minute dots in varied tones, creating a coarse, gritty, and washed effect. Within this field, small black specks, a few millimetres across, appear sporadically, resembling sunspots seen through a telescope. At the edges, faint scratches and abrasions suggest traces of human handling, and in all four corners, small black punctures, like staple holes, evoke the sense of a document weathered by time and use. Along the left edge, near the border, pale yellow sans-serif lettering, about a centimetre high, reads: 'FUJI 14846 AEBIEH (20.3 × 25.4)'. If one looks at the information plaque next to the artwork, they will find the title and description of the piece:

If the Radiance of a Thousand Suns Were to Burst at Once into the Sky
2015 Enlarged and dry-mounted print of an 8×10 photographic film buried for three months in a location where the radiation level was 34 μSv/h

Post-catastrophe Society in Japan

It was June 2013 when I first entered the no-return zone of Fukushima, officially known as the 'Difficult-to-Return Zone' (DRZ). More than two years had passed since March 11, 2011 earthquake that struck Japan's north-eastern coast, triggering the meltdown at the Fukushima Daiichi[1] nuclear power plant. The magnitude-nine quake not only caused immense destruction but also forced mass evacuations within a 20-km radius of the plant, displacing thousands permanently (Kushida 2012).

During those 2 years, my understanding of the nuclear accident, and of the issues surrounding it, gradually evolved, much like society's perspective on the problem. In the first month after the earthquake, there was a period of societal shock and subdued panic, not always visible but certainly deeply felt. Volunteers mobilised across tsunami-hit regions, supporting evacuations, logistics, and shelters. In the media and across social networks, one could sense an atmosphere of solidarity, albeit somewhat oddly formal, even aggressively polite. In this initial phase, many visual artists responded swiftly and energetically to the nuclear accident in Fukushima. For example, the artist collective Chim↑Pom created the artwork *Kiai Hyaku Renpatsu* (Chim↑Pom 2011), a 10-minute video of participants shouting 100 slogans in Soma City. They also initiated the collective project *Don't Follow the Wind*, together with curators Kenji Kubota, Eva and Franco Mattes, and Jason Waite, in which a group of international artists created artworks that would remain present but unviewable within the DRZ until public access is restored. Other artists such as Kyun-Chome, Kota Takeuchi, and Bontaro Dokuyama also addressed these issues with equally critical engagement.[2]

I, however, found myself unable to respond artistically. I confined myself to volunteer work or photographic projects, without any intention of public exhibition. Even though I knew that making artworks that explore the different aspects of these issues might help raise awareness and open space for dialogue, I also felt that putting my authorship on something derived from such suffering might be a kind of an amoral complicity, a form of hypocritical artistic appropriation. This hesitation only intensified as I watched how polarised and emotional the public responses to controversial artworks could be. Instead of dialogue, many of these moments became echo chambers of outrage and self-affirmation. As a result, the only thing I seemed capable of producing was restrained inaction.

Another factor that added to my reluctance was the staggering pace of the mass and social media. I believed the issue of nuclear energy dependency had far-reaching consequences, both geographically and temporally. Yet, despite its long-term impact, the media, along with social, political, and artistic activity, seemed to focus on the immediate, sensational aspects of the disaster. These responses were powerful but quickly consumed and replaced by the next current of activity. I felt that a slower form of language was needed, one that could complement daily information and address the apparent void in medium and long-term engagement.

Over the following years, social attention gradually and politely shifted away from the critical issues, and the visibility of the problem gradually faded. However, in June 2013, during an artist-in-residence at Tokyo Arts and

Space (TOKAS),[3] conversations with fellow artists gradually shaped the idea of researching the DRZ.

How to Enter the Difficult-to-Return Zone of Fukushima

Most logistical preparations are relatively simple: acquiring N95 masks, hazmat suits, shoe covers, wet wipes, latex gloves, and related equipment is straightforward. The more critical task lies in assessing risk and determining appropriate readiness. Radiation penetrates most fabrics, making external exposure unavoidable. For this reason, carrying a Geiger counter is essential to monitor exposure and calculate permissible doses according to public safety standards.

As of 2014, the annual radiation dose limit for the general public, defined by the International Commission on Radiological Protection, was 1 millisievert (mSv) per year (Valentin and International Commission on Radiological Protection 2007). In emergencies, this threshold could rise to 20 mSv. In Tokyo, background radiation was about 0.04 microsieverts (μSv) per hour, equating to roughly 350-μSv annually, far below the 1,000 μSv (or 1 mSv) limit. By contrast, at 10 μSv per hour, a radiation level that I encountered in some hotspots along the DMZ, one would reach the annual public dose in just 100 hours.

The most dangerous form of contamination is internal radiation exposure, which occurs when radioactive particles like caesium-137 are inhaled. These particles continue emitting radiation indefinitely within the body, even after leaving contaminated areas. Hazmat suits and N95 masks primarily prevent radioactive dust from adhering to clothing or entering vehicles where it might later be inhaled. Therefore, maintaining strict protocols that separate contaminated exterior zones from clean interior spaces becomes vital, keeping all traces of earth or dust out. Since radiation remains largely invisible, this procedure must be followed systematically and rigorously to confront contamination risk and enable work to proceed with minimal psychological burden and avoid panic.

There are several ways of entering the DRZ. The first, and more demanding, is through participation in reconstruction or decontamination work. These efforts have continued for over a decade. Due to legal limits on radiation exposure, there is a persistent shortage of manual labour. Some artists, including Kota Takeuchi's collaborator in the *Finger Pointing Worker* (Finger Pointing Worker and Takeuchi 2011), or Masaru Iwai in his *Control Diaries Project* (Masaru Iwai 2020), have used this approach as a means of accessing the zone. The second route is to obtain a permit, typically granted by the authorities of the affected municipalities or the specific area one intends to visit. The boundaries of the zone have shifted over time, gradually shrinking, though not dramatically. As March 31, 2025, the area encompasses parts of the following towns and municipalities: Iidate Village, Minamisōma City, Namie Town, Katsurao Village, Futaba Town, Ōkuma Town, and Tomioka Town. The nuclear plant is located on the border between Futaba and Ōkuma. Again here, there are essentially two categories of entry permit: one for residents and another for activities deemed in the public interest. The first type of permit is issued to displaced residents requiring temporary access to their original homes. This permit allows them to be accompanied by others, meaning any evacuated resident may host visitors in the area. The second category is available to anyone, provided the visit serves a public-interest purpose. On my first entry into the zone, I used this type of permit, which the staff at the TOKAS artist residency, as a public institution, kindly offered to arrange. On later visits, I obtained the same permit through personal application as a photographer or artist.

First Contact

On June 3, 2013, after renting a car in Tokyo and preparing my luggage, camera, recorder, protective suits, and documents, I leave the city at a 14:45 in the afternoon, aiming to reach the affected area and Fukushima by 17:30. After exiting the motorway and joining National Route Six heading north, the surroundings gradually begin to change. Shops and houses lining both sides of the road start to appear abandoned, and little by little, the landscape takes on an air of desolation. Before reaching the DRZ, I stop once to photograph some fields filled with enormous black bags, which contain surface layers of soil that have been collected and relocated during the decontamination purposes. A police patrol car pulls up next to mine, and two officers step out to ask a few questions. After I show them my permit, they wish me a good day and drive off almost immediately. The radiation level on the Geiger counter reads 0.45 μSv per hour. Once I complete the photographs and documentation, I head to the hotel I have booked in Minamisōma to prepare for entering the zone the following day.

The next morning begins sunny, and fresh. After leaving the hotel, I drive to the nearest convenience store to buy water, and put on the hazmat suit and mask that I will not remove during my entire time inside. Naturally, there are no shops or toilets within the zone, as removing one's mask or protective suit would be far too hazardous. The perimeter is fully fenced off, allowing access only through four or five official checkpoints. Upon arrival at one of these, I join a short queue of about five or six contractor vans and lorries. While I wait, I watch as the checkpoint workers, all wearing masks and protective clothing, carry out their tasks in a disciplined and rhythmic sequence. Stop, check, wave through. Stop, check, wave through. Everything is handled with efficiency and courtesy. When it is my turn, I present my permit along with my personal documents. One of the workers confirms the car's number plate and reviews the paperwork. He greets me with a friendly 'Otsukaresama desu',[4] and I am asked to move in. In this way, I find myself inside the DRZ. Most vehicles continue directly to the reconstruction works at the nuclear power plant, but I take a detour and begin exploring the towns of Futaba and Ōkuma with a strange sense of anticipation, uncertain what I might encounter.

The scenery appears familiar, yet eerily unknown, almost overly deferential, as though one has stepped into a mirrored version of the known world. In the town, houses half-destroyed by the earthquake two years earlier remain untouched, neglected, shrouded in dust and overgrown with vegetation, like remnants from a post-apocalyptic science fiction film. The traffic lights, all switched off and silent, seem to amplify the sensation of suspension—of traffic rules, civil logic, and time itself brought to a halt. The overall atmosphere is one of profound and haunting desolation. Nature, with quiet determination, is reclaiming what has once been hers. As I move slowly through the towns, occasionally stepping out of the car to explore or photograph, a young wild boar approaches to within several metres and observes me curiously, showing no sign of fear. I stand still and watch. After a few moments, the boar loses interest and trots off, vanishing into the remains of a partially collapsed house. It has likely never seen a human before and shows neither alarm nor caution for long.

As one leaves the towns and enters the blurred rural zones and little pine forests, some roads are cracked, and others are overtaken by foliage, causing many to lead to dead ends and forcing detours. Inside the car, the radiation increases, and the shrill beeping of the Geiger counter rises and falls, oscillating between 2 and 10 μSv per hour. Outside the car, beyond the Geiger counter's continuous beeping, only bird calls from the forest and the far-off sound of repair machinery at the nuclear plant disturb the quiet.

At around 14:00, I leave the zone again and head to the nearest Seven Eleven to have a quick lunch and use the bathroom. After a short rest in the car park, I return to the zone. This time, I head towards the entrance of the nuclear plant. In the afternoon, a thick fog, like smoke or dense smog, has arrived with the wind from the southeast. It is unsettling and eerie. In the surrounding area, radiation levels are relatively high, sometimes exceeding 10 μSv.

In the large outer car park, many empty buses are parked, waiting for the workers to finish their shifts. Amid the mist and the ghostly outlines of chimneys with their constantly blinking lights, the shadows of the square-shaped reactors can be seen. I explore the area a little and, finding a discreet spot, I take my large format camera out of the car to begin photographing. The process is slow and tedious because of the gloves, mask, hazmat suit, and full safety protocol required for radiation protection. Behind me, I hear several groups of workers approaching, returning from the outer zone to the plant. When I turn, I see two groups with shovels coming towards me, and I begin to prepare in case they tell me, as has happened before, that photographing the nuclear plant is not allowed. To my surprise, each group passes by my side, looking with open curiosity at my antique wooden camera. One of them, in an animated tone, asks, 'What are you photographing?' and they continue towards the plant entrance. The workers have their surnames written in large hand-drawn letters on the backs of their white hazmat suits, since with the masks and the identical white suits, it is impossible to tell people apart. After they pass, I continue taking more photographs, filming, and recording sound. Because of my hazmat suit, the same one worn by everyone in the zone, no one seems to suspect or pay any attention to my activities. In fact, the anonymity created by the suits and masks fosters a kind of camaraderie among the workers, the only people permitted inside the zone, and I become one more invisible figure within this space.

Shortly after, at five o'clock in the afternoon, multitudes of workers begin to exit the plant, each heading to their respective buses. Curiously, on the front windscreen of each bus there is a sign with a number and, written in large letters, simply the word 'Hotel', 'Izakaya' (bar tavern), or 'Pachinko' (a Japanese form of gambling game often played in dedicated halls in rural Japan). Each time a bus fills, the doors close and it departs, heading out of the zone. I also finish photographing and head to the checkpoint, then make my way to the North screening point, the closest one. The screening points are places where a team of three or four employees with radiation monitors check the exterior of each car, the wheels, and then the body of each person, with special attention paid to the soles of the shoes and any equipment that has touched the ground, to ensure that no radioactive particles have adhered. Here, one can also dispose of used suits, gloves, shoe covers, and masks.

The hotel where I am staying is a modest workers' hotel north of the zone in Minamisōma. Like the night before, it is full. My room is in a newly built wing that looks temporary and cheap, constructed with prefabricated thin walls. Since the 2011 accident, such hotels have proliferated in towns bordering the zone, though even now they do not suffice. Major general contractors managing reconstruction often rent out entire hotels during the week to house the temporary workers. At nightfall, in the hotel dining room where both dinner and breakfast are served, about a dozen male workers, ranging in age from around thirty to fifty, eat their meals while watching a baseball match on television. Most sit alone and few speak. Some sit with a can of beer or strong alcohol, available from a vending machine in the hotel lobby. A few, already worn out from the day's labour, have a towel draped around their necks, heads bowed, and hands resting on their beer, dozing off. In general, the air is heavy with dusty resignation and fatigue.

At about 9 o'clock, a minibus stops in front of the hotel and a louder group of workers enters. Most have flushed faces and appear to have been drinking. After chatting excitedly for a few minutes at the hotel entrance, they say their goodbyes and disperse. Something I came to learn later is that the bars, cabarets, and all the nightlife establishments in these peripheral towns are undergoing an economic boom never before seen. Temporary workers from all over Japan are drawn by the work and relatively good pay of the decontamination and construction projects within the no return zone. They immerse themselves in the daily grind of manual labour and well-paid temporary tedium for a few months. To alleviate the solitude, the monotony, or the unbearable nature of the situation, many of these male workers fill the local bars after work, like a swarm of moths drawn to light in the night.

Over the following 2 days, I continued exploring, focusing more on the abandoned corners of the no return zone, which most fittingly resembled 'the Zone' from the brilliant and almost prophetic film *Stalker* by Andrei Tarkovsky, a place of paradoxes where the logic of daily life loses its meaning. It is a place that is both near and far at once. The no-return zone lies less than three hours from Tokyo, yet it remains one of the most restricted, remote, and isolated areas in Japan. Within a maximum radius of 20 km, no one resides. Only reconstruction workers systematically enter in the morning, leave for lunch, return in the afternoon, and finally exit the area at the end of their working day. After nightfall, not a single light shines from the windows of the abandoned houses. In the place where much of the electricity that once powered the lights of Tokyo was produced, only the most desolate darkness now reigns. 'The aura is the appearance of distance, however close the thing that calls it forth', writes Walter Benjamin in *The Arcades Project* (Benjamin 1999, p. 447). In this case, created by the paradoxical proximity and simultaneous remoteness of the no return zone, the aura it emits is one of darkness; like a black hole, a negative aura that stands out on the map at night by its absence of light.

Return to the No-Return Zone

In September 2013, three months later, I return to the area. On this occasion, I decide to take photographs and document the interior landscape, already estimated to be inaccessible to the public for at least 100 years. I am interested in finding a way to fight against oblivion. By mid-September, the weather is sunny and hot. The lush green trees, almost foamy, stand tall in the distance, proud, with unusual insolence, forming a kind of distant wall, as if constructing the end of the world, the boundary beyond which there is only vast, transparent, lucid blue sky. The immediacy of the sky is penetrating too. Fukushima shares its latitude with Seville in Spain, or San Francisco on the American continent. This positioning means the sun's rays strike the earth's surface at a greater angle, something noticeable in both the strength of the colours and the reflection angle of light. The chatting of birds, sparrows, and various other species is always pierced by the obnoxious beeping of the Geiger counter alarm. These noisy beeps seem to pull back part of my mind, never allowing full surrender to the environment, as if that alarm sound were the only thread connecting you to the world of sanity. Once you surrender to the false sweetness of nature, to the soothing sound of distant waves, you risk being lost forever in a fabricated Eden. Fabricated, however, only for the humans who built it. The animals, nature, the air, the earth, and stones continue their course, indifferent to the radioactive future surrounding them. In that sense, it is sublime space, distant, strict, primitive, primordial and aureous in essence. Like an earthly utopia, in the literal sense of the Latin word U-topos ('no-place').

During this trip I, choose to experiment with documenting the area using 8×10 inches large-format photographs. Instead of bringing the films back and developing them, I opt to leave some buried in the ground to be collected later, or possibly to remain there permanently for years, until access becomes possible again. Divided across various locations with differing radiation levels, I begin photographing and burying films at each site. As I engage in the physical act of digging earth with gloves, masks, and hazmat suits, I become acutely aware of the materiality and fragility of the human body. The suffocating conditions, compounded by the physical exertion of digging and the constant need to wear protective gear, makes the experience almost nauseating. My clothes beneath the suit are soaked with sweat. This physicality, ironically, enhances a sense of total rejection. Nothing can be touched directly, nothing can be smelled. The only scent I have experienced there is the industrial and highly plastic odour of the interior of my N95 mask. Effectively, it is a location that actively resists human presence, persistently and abrasively. The only form of interaction permitted is indirect, always mediated by a thin layer of plastic or rubber. Hands touch everything through latex gloves. The area is present, but never fully accessible. Even the natural sounds, the waves, the wind, and the rustling trees are pierced by the shrill, sordid ringing of the Geiger counter. And most disconcertingly, the radiation and the danger that the Geiger counter alerts us from is not visible; it is again slippery and visually inaccessible. Here, every action or movement demands total concentration, transforming each task into something unfamiliar and entirely strange.

Simultaneously, with the immediacy of nature, the total absence of human commercial activity becomes disturbingly intense. A metaphorical and literal sense of being offline accentuates this strange void, a lack of stimulation, a lack of excess. According to Georges Bataille (2007), we live in a society of excess, where the wasting of this excess of resources is associated with 'luxury' and characterises our current society of overconsumption[5], but inside the no-return zone, the absence of this excess becomes absolute.

The nuclear energy industry represents the apex of this philosophy of excess. The volumetric energy contained in matter varies depending on the material. For instance, nuclear fuels possess a volumetric energy density that is tens of thousands of times higher than chemical fuels. A uranium fuel pellet the size of an inch is equivalent to one ton of coal, 120 gallons of crude oil, or 17,000 square feet of natural gas (USDOE Office of Nuclear Energy 2019). This abundance creates redundancy – whether energetic, material, or psychological – which in turn fosters a sensory atrophy that Susan Sontag describes in *Against Interpretation* (Sontag 1966). In our capitalist society structured by efficiency and consumerism, these consequences are concealed, so no visible contradiction hinders economic activity and consumption. One of capitalism's impregnabilities is that we, as consumers, are both victims and perpetrators. When daily decisions are driven by economic logic, we relegate our value systems to the background. We replace qualitative personal worth with quantitative economic value. Capitalism tries to make us believe the two are interchangeable. What makes this tension nearly invisible is the complex logistical machinery that conceals internal contradictions. But inside the DRZ, we can see the result. Here, the cost of the electricity we consume, and sometimes waste, becomes starkly visible. These sites, typically located in remote regions, function like classified documents: public in theory, yet practically inaccessible. Inside the DRZ, all these negative by-products condense into a tangible presence. The absence that remains latent in broader society becomes, in this place, manifest and real, reciprocal to the conspicuous lack of commercial excess in the DRZ.

Coda

Over the year, I entered the zone four times. In April, after developing several of the 8 × 10 negative and positive films I had left buried, I was finally able to observe the experimental outcomes. The first photographs I had taken in September and buried showed significant contrast problems. They lacked depth and blackness, as though the film had been accidentally exposed to light before being used. I also discussed the results with experts at the developing laboratory. The outer edges of the film, which are usually black because they are covered by the holder during exposure, appeared grey and pale instead. The technicians suggested that the issue was unlikely to be accidental, as it did not resemble a typical photographic error. Rather, it seemed chemical, as if something had penetrated the film holders. I then began experimenting with unexposed film to further understand the phenomenon.

The final negative I collected in March, which had remained sealed and unexposed throughout the burial period in the irradiated zone, returned from the laboratory showing an orange, transparent, and radiant tint. It appeared as though it had been exposed to sunlight. Upon first seeing the negative, I was absorbed in awe at the colour's aesthetics, but this was swiftly followed by guilt and contrition. Once again, the familiar dilemma between the tyranny of authorship and the aesthetics of morality began to resonate within me. After several weeks of hesitation, I decided to present the work at the end-of-residency exhibition, alongside other projects relating to the nuclear crisis in Tokyo. After all, this may not have been photography in the conventional sense. Perhaps I was not even the author. In her article *Radical Contact Prints*, Susan Schuppli writes about Vladimir Shevchenko's film, who visited the Chernobyl site three days after the meltdown in April 1986, to find his developed 35-mm film was full of noise and interferences, and explains that

Shevchenko eventually realised he had captured 'the image and sound of radioactivity itself' (Schuppli 2015, p. 287). Schuppli argues that 'the radioactive comes into visibility when representation is no longer the objective of image-making and the making of the image itself becomes the event' (Schuppli 2015, p. 289). For her, the 'radical contact prints' are not the indexical representations of a signified event, but they become signifiers by themselves. Building on Schuppli's lucid argument, Atsu Ito further writes in *The Spectral Image: The Brief History of Visualising Radiation* (Ito 2018, pp. 2–3):

> [...] while a photographer or a filmmaker tries to represent the scene of irradiation, they become a representation of this invisible energy. Thus, the operator of the apparatus of capture becomes captured as an indexical representation of radiation. In a paradoxical way, the human subject who attempts to capture a form of energy to translate it into visuality becomes the outcome of the process.

In this sense, it could be argued that the actual work was not a referential photograph taken by me but evidence of a process in which I, the subject, was transformed into an object of radiation, a subject-object. That, perhaps, was a trace, not of me producing the exposed films, but the films operating on me, pushing and pulling me to create an irradiated human object out of me.

A year later, the work was nominated for an annual exhibition at the Ueno Royal Museum in Tokyo and was selected to form part of the collection of the Ohara Museum of Art in Kurashiki. Since then, it has hung quietly on the white walls of the gallery, a luminous field of orange colour, suspended behind Plexiglas, where its smooth, reflective surface continues to absorb and return the gaze of those who stand before it, bearing silent witness to the encounter between radiation, material, and body.

Notes

1 Meaning 'first or number one' in Japanese.

2 For a more complete overview of artistic responses to the catastrophe, see Chapter 8 of *Invisible Colours* (Decamous 2018), *The Nuclear Culture Source book* (Carpenter (ed.) 2016), or the article *Revealing the Radioactive Contamination after Fukushima in Japanese Photography* (Davre 2019), among other references.

3 Formerly known as Tokyo Wonder Site or TWS.

4 'Otsukaresama desu' is a common Japanese expression used among colleagues, used in appreciation for someone's hard work and effort which literally means something like 'your respectful tiredness'.

5 See Bataille (2007). And in relation to this, Sontag argues that precisely it is this excess which dulls our sensory perception (Sontag 1966)

References

Bataille, Georges. 2007. *The Accursed Share 1: Consumption*. Translated by Robert Hurley. New York: Zone Books

Benjamin, Walter. 1999. *The Arcades Project*. Cambridge, MA: Belknap Press

Carpenter, Ele. (ed.)., 2016. *The Nuclear Culture Source Book*. London: Black Dog Press

CHIM↑POM, 2011. *Kiai Hyaku Renpatsu*, 10:29 minutes, single channel video, color, sound. Collection of the Mori Art Museum, Tokyo.

Davre, Amandine, 2019. 'Revealing the Radioactive Contamination after Fukushima in Japanese Photography'. *Trans Asia Photography*, 10(1). Available at: https://read.dukeupress.edu/trans-asia-photography/article/doi/10.1215/215820251_10-1-104/312733/Revealing-the-Radioactive-Contamination-after. [Accessed 7/25].

Decamous, Gabrielle. 2018. *Invisible Colors: The Arts of the Atomic Age*. Cambridge, MA: MIT Press.

Finger Pointing Worker and Takeuchi. 2011. *Pointing at Fukuichi Live Cam*, 24:40 minutes, Single-channel video, color, sound. Represented by Kota Takeuchi. Collection of the Kadist Collection, Paris.

Ito, Atsuhide. 2018. 'The Spectral Image: The Brief History of Visualising Radiation'. *The International Journal of the Image*, 9(2), 1–5.

Iwai, Masaru. 2020. *Control Diaries*. Tokyo: Takuro Someva Contemporary Art.

Kushida, Kenji. 2012. Japan's Fukushima Nuclear Disaster: Narrative, Analysis, Recommendations. Stanford University Shorenstein Asia-Pacific Research Center Working Paper, Available at SSRN: https://ssrn.com/abstract=2118876 or http://dx.doi.org/10.2139/ssrn.2118876.

Schuppli, Susan. 2015. 'Radical Contact Prints'. In John O'Brien (ed.), *Camera Atomica*. London: Black Dog Publishing, pp. 277–291.

Sontag, Susan. 1966. *Against Interpretation, and Other Essays*. New York: Farrar, Straus & Giroux.

USDOE Office of Nuclear Energy. 2019. *The Ultimate Fast Facts Guide to Nuclear Energy*. No. DOE/NE-0150.

Valentin, Jack and the International Commission on Radiological Protection (eds.). 2007. *The 2007 Recommendations of the International Commission on Radiological Protection*. Oxford: Elsevier.

3 The Nuclear Archive

Daniel Beck

In 2020, I began a practice-based PhD focused on sound and nuclear culture at CRiSAP (Creative Research in Sound Arts Practice) at the University of the Arts London, which I completed in 2024. My research included an extensive sonic fieldwork at the British Film Institute's (BFI) National Film and Television Archive site in Gaydon, Warwickshire. The BFI's facilities at Gaydon are situated within a former Cold War era nuclear weapons depository that once served the nearby RAF Gaydon and the V-Force era bombers that were equipped to carry the UK's first nuclear bombs.[1] The purpose of this fieldwork was to explore how acts of listening, remote sensing, and archival research might be equipped to inherit a kind of 'atomicity', where nuclear cultural spaces and their artefacts can be interpreted by attuning to their atomic consequences and what they represent.[2] Utilising practices of documentation, sonic experimentation, and film media at the archive site, I produced a series of interconnected artworks that skirt the boundary edge of where this space of media preservation coexists with the enduring nuclearity of the buildings, bunkers, and lands surrounding the site. Through this process of sonic experimentation, I conducted a search for the ways in which sound might commune with the atomic and enable practices that can challenge the limits of sonic-atmospheric and sonic-energetic encounters.

During this process I produced a series of multi-format artworks, including a web-based work, titled *Radionuclide Websites* (2022-present) and a 35-mm film project, titled *The Nuclear Archive* (2022). *The Nuclear Archive* is a fifteen-minute art film that includes footage, field recordings, and the sound and images from *The Radionuclide Websites*, which consists of a series of 'digital radionuclides' – web-based representations of nuclear materials. The film, exhibited at the 'Radiant Objects' exhibition in 2023,[3] documents the atmospheric and sonic engagements captured as I exposed my digital radionuclide materials through a series of media transductions that I undertook in partnership with the BFI's archivists at the National Film and Television Archive throughout 2021–2022. The film also includes sound and images from Tabloo, a visitor centre on the site of a planned nuclear waste depository

DOI: 10.4324/9781003622185-4

in Dessel, Belgium. What emerges through the documentation of my practice at the BFI and at Tabloo is a tangled sonic thread, which is constructed from the documentation of the post-atomic, atmospheric, and cultural artefacts that surround the BFI archive and Tabloo sites. The audio track includes the sonic architectures of former Cold War era military structures, magnetic tape machines, optical audio tracks, digital glitches, and the hum of electronic circuitry. It is the substance of the combination of these sounds that I argue contributes to the generation of a 'noise-print' that is unique to this nuclear cultural archive. In this chapter, I will explore how my practice at the BFI uncovers the ways in which sound and atmospheric conditions can interact within sites of nuclear significance and make audible their nuclear cultural consequences.

Post-Atomic Auralities

There are many ways in which artists have developed and engaged in art practices that highlight and follow the interlinking strands of nuclear effects and their consequences. In an essay titled 'Trace Evidence: A Nuclear Trilogy' (2016), the artist and author Susan Schuppli highlights an example of this within her own work by detailing the various conditions that led to the detection of radioactive materials from Chernobyl at a nuclear power station in Sweden, before the announcement of the disaster in Chernobyl's neighbouring town of Pripyat (2016, pp. 37–42). For Schuppli, the emergence of this radioactive material in Sweden is a 'material witness that would first come to testify publicly to the accident of Chernobyl' (2016, p. 40). In addition to bearing witness in this way, for Schuppli, these materials are also representative of the time-delayed announcement of the disaster. The discovered material is more than a witness to the radioactive effects of Chernobyl, it is also witness to the many cultural, political, and social effects that proliferate as a consequence of the event – such as the decisions of politicians and officials to choose when to inform the public of the immediate danger to their environment and health. As Schuppli states in her description of her video artwork *Trace Evidence* (2016), 'Despite its radical and covert nature, the unique signature and behaviour of radioactive isotopes allows its lethal traces to be tracked directly back to their source, re-connecting the evidential links that planetary phenomena has seemingly torn apart' (2025). Schuppli's material witness is an example of how the 'atomicity' of our nuclear actions can be measured by their cultural, political, and societal impacts as well as their direct material-radioactive effects. Where the presence of radioactivity is often both hard to detect and potentially dangerous, the enduring effects of the post-atomic can be seen, heard, felt, and experienced in more ways than the radioactive emissions of the materials they generate. What happens when the *always*-effects of the nuclear are traced back not to their lethal ionising effects but to their cultural by-products? Where do these begin and end, and how might we replicate, intercept, or learn from them?

The overriding purpose of nuclear culture studies is to investigate the enduring effects of nuclear activities, irradiated landscapes, and materials, while placing them in the context of their associated cultural transformations and influences. Studying interactions with irradiated objects, lands, and fissile materials provides a reference point from which we can learn from their 'radiant' effects. It is understandable that much of our knowledge gathering on the cultural effects of ionising radiation comes from the historical accounts of nuclear incidents and the violence of nuclear explosions. Further to this, nuclear cultural studies can reach into the medical records of those affected by nuclear events (Kuchinskaya 2023), or include the dosimeter clicks of ethnographers, scientists, and artists whose practices come with cautious interventions and warnings of the *always*-effects of our nuclear actions.[4] Tethered to the reliable timepiece of the half-lives of heavy metals and manmade radioisotopes, nuclear cultural discourse has established the 'radiant object' that this book seeks out as a 'super-enduring' material manifestation of the atomic era. This 'radiant object' is a geologic marker whose millennial durational effects are entwined with cultural influences, technological developments, and environmental consequences. Within my research, I have utilised a sound-based practice that considers what we might learn from tuning in to the threads that connect nuclear materials, media, cultures, and events. It is an approach that considers sound as an active component in the story of the nuclear, where the audible is engaged as both a wayfinding instrument and a carrier of the post-atomic.

The Digital Radionuclide

Radionuclide Websites (2022–present) is a series of eight connected webpages that are designed to emit audible and visual fluctuations throughout a lifecycle they share with particular nuclear materials: Cm-244,[5] Kr-85,[6] Pu-238,[7] Pu-240,[8] Pu-242,[9] Sr-90,[10] Cs-134,[11] and Cs-137.[12] In the context of the timelines of our interactions with nuclear materials, the *Radionuclide Websites* deposit themselves as a digital site of radioactive effects. Each radioactive-(web) site's date of expiry is locked to the half-life of the material they represent. As with the decay of radioactive materials, this process is continually generative, where the process of nuclear decay is transposed to the transmission of colour and audible frequency over time. The digital radionuclide acknowledges the digital space as a construct of complex systems that includes energy networks, programming, and the physical materials that make up the core components of servers, computers, and personal devices. It is material and inseparable from the atmospheric-climactic effects of fossil fuels, and the geologic and cultural impacts of the extraction of materials, which includes the uranium used in nuclear energy production. The survival of the digital radionuclide is highly dependent on the fragile nature of this complex network, and this directly impacts their ability to be found, accessed, and remembered in the future.

The *Radionuclide Websites* started as a digital experiment, a testing of the boundaries of 'atomicity', where a measure of atomic burden could be re-imagined as a post-atomic-atmospheric, sonic, or cultural effect. However, as the project developed, the digital radionuclide began to take on new and unexpected shapes and forms. While observing the scale and impact of their inherited nuclearity, I was aware that I had placed these new digital-nuclear materials within a space of languages, technologies, and programming that are in a permanent state of flux. Ascribed with a nuclear-millennial timeline, these radionuclide websites share similar vulnerabilities and challenges that arise when considering the long-term storage of their nuclear material counterparts. During the development of *Radionuclide Websites,* I had discovered the nuclear history of the BFI archive site in Gaydon and had begun an extensive sonic fieldwork aimed at uncovering the nuclear sonic potential of the bunkers and grounds surrounding the site. This research included the production of a detailed sound map of the site, which includes interviews with the BFI archivists and a former Cold War era nuclear armourer.[13] Within the sound map, the descriptions of the delicate processes involved in handling both nuclear and film materials reveal similarities that merge the modern media archive with its nuclear cultural history. Considering the digital radionuclide as a fragile decaying medium with an 'atomicity' of sound and image opened up further avenues for exploring how these digital-nuclear materials might change and adapt when exposed to the archive space.

Nuclear Sonic Fieldwork

In 2021–2022, I conducted a series of on-site experiments at the BFI archive site in Gaydon where the *Radionuclide Websites* were transcoded, played back, and re-recorded from digital video to 35-mm polyester film, and magnetic tape. This experimentation transformed the source material, re-setting their timeline of expiry to that of the transcoded physical media, while also inheriting the value of their newly acquired material form. This transduction of web-based work to film material initiates a transformation of the *Radionuclide Websites* from a medium of transmission to a recorded format. In *Decomposed: The Political Ecology of Music* (2019), media theorist and author Kyle Devine approaches the subject of the materiality of music media, stating that 'media of transmission overcome space but cost time, while media of recording overcome time but cost space' (Devine 2019, p. 152). In the live auditioning of the digital radionuclide's sound and image at the site of the BFI's archive, the balance between what is recorded material and transmission is troubled. The recorded sound captures the glitches of videotape technology, the cracks and pops of 35-mm film devices, and the sonic architectures of the bunkers, grounds, and archive buildings that contain them. This process of audition and transduction reveals how recorded media can also start to collect nuclear cultural and technological markers that can be made traceable to their long-term effects. Within the context of sound recording, I describe these markers as 'noise-prints'. Noise-prints provide aural references for sound arts practitioners that are unique to their point of capture – they can be recorded and translated between archives, text, film media, and audio recording devices. Noise-prints are aural evocations of the environments they communicate with, the atmospheres they share, and the cultures surrounding them, and it is this that I argue makes them a suitable carrier for post-atomic cultural effects. Within my practice, I listen for how noise-prints might be captured, transported, and reactivated through the sharing or playback of the media and materials that contain them. *The Nuclear Archive* provides one example of how a noise-print can be captured and attributed to two sites of nuclear significance.

The Nuclear Archive presents each location, transposition, and activation of the *Radionuclide Website* materials during my research. As well as containing footage of my activities across the various BFI archive sites, it also includes sound and video that I gathered during my attendance at a week-long conference in Dessel, Belgium. The sound in the film overlays and undulates between different versions of the same soundtrack, one is the playback of the digital radionuclides through the BFI's Steenbeck at Gaydon,[14] a VHS transfer of the same film, and the original digital audio capture. As the camera focusses on the turrets and corrugated steel walls obscured by the forests surrounding the Dessel nuclear storage facility, we hear the whirring turntables and crackling speakers of the Steenbeck. At the halfway point in the film, the full recorded cycle of the radionuclide websites transitions from a framed shot of the VHS monitor to the original digital file, while their respective soundtracks are transposed in the opposite direction. This displacement of the sonic and visual representations of the radionuclide materials draws out and places emphasis on their differences and the transformation that has occurred during my archival study. As the noise-print of the archive replaces the shrill high-pitched frequencies of the digital radionuclides, the image becomes increasingly erratic, confined to a fixed-screen-width, the density of the waveforms in the *Radionuclide Websites* algorithm comes in and out of phase, generating a complex shape-shifting pattern of flashing colours that elicits imaginaries of nuclear exposure.

The clarity of the audio from the digital file at the start of this sequence evokes the slow ramping frequencies of air raid sirens. As the digital audio file is gradually transitioned to the VHS tape recording, audible artefacts from the playback of the film within a bunker on site begin to be introduced, and the noise and atmospherics of the bunker space increasingly interject

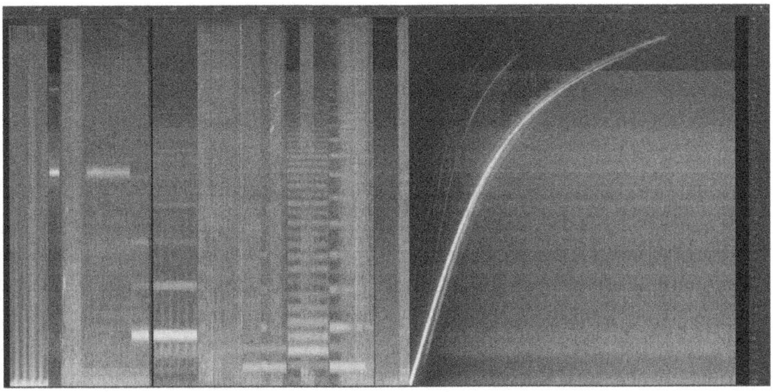

Figure 3.1 This sonograph image of the soundtrack from *The Nuclear Archive* shows the intensity of particular frequencies within the soundtrack overtime. The large curve seen on the right side of the image shows the ramping of frequencies as the film accelerates through the programmed sequence of the *Radionuclide Websites*. Image © Daniel Beck (2024).

and become more prominent. The effect of this is visualised in the sonograph image (Figure 3.1) by the curve on the right of the image that shows a bright white stripe.[15] Beginning at the lowest frequencies, the image is clearly defined and uninterrupted by other noise or sounds, which are represented by the grey area that becomes more prominent as the transition between the soundtracks develops. As the range of recorded frequencies increases beyond the capabilities of the individual listener's ability to perceive them, a new sound emerges. It is audible within the film's soundtrack from around 12 min 50 s, although this may be more or less noticeable at different times depending on the listener – a new ramping of frequencies can be heard, cycling in pitch throughout the remaining recording. This warped audio effect is not present in the digital file, and it is unclear when this sound first appears in the recorded media, but as with the unexpected complexity of the waveform visualiser's algorithm, it is a unique imprint entirely of the technology's making. These sonic reflections can be seen in the sonograph image as light scratches; they are aural imprints that are unique to the noise-print constructed through my practice and are both a marker and carrier of the post-atomic within *The Nuclear Archive*. Transposed to film the soundtrack can be cut-up, shared, and distributed as a radiant object whose 'atomicity' is carried not in the direct threat of radioactive consequences but as a 'radiant archival media'. Stored within film cans, the film enters the archive as a point of record for the long-term durations of the *Radionuclide Websites* that is also infused with the post-atomic-sonic architectures of the BFI archive's unique nuclear cultural history.

Conclusion

In combination with the archived materials at the BFI, my practice is an exploration of the complexities that occur when processes of audition (playback), distribution, and sound gathering are attuned to nuclear cultures, as well as those of sound and film practice. It is in an observation and sensitisation to the shifting narratives that occur between archive, audience, and practice that I have situated my concept of noise-prints as a new approach for engaging with(in) archives. Activated through sound arts practice, a noise-print can both interpret and document aural, atmospheric, and cultural effects. It is a practice that utilises data, documentation, and recording techniques to capture an aural representation of nuclear events and places. The long-term future of the content of the BFI archive is secured through processes such as digitisation, duplication, and transference to other film media. However, preservation of the physical material media – acetate, polyester, and the highly flammable cellulose nitrate substrates and their audio tracks – is also considered vital to the archivists and technologists that work on the collection. The Master Film Store located at the BFI facility in Gaydon is a world-leading archive building designed to keep the most sensitive and volatile of these materials in cold stasis, preserving not just their film and televisual content but also the substance of their material formats. In addition to this modern cultural-material archive, the BFI utilises and maintains a series of Cold War era nuclear bunkers, their associated workshops, and other outbuildings that are buried beneath mature trees, mosses, and bramble throughout the grounds. *The Nuclear Archive* uncovers the unique aural signature, or 'noise-print' of these spaces and activates them as a nuclear cultural marker.

In its physical film format, *The Nuclear Archive* exists as a representation of a 'radiant object' that presents the variations, artefacts, and technologies of the archive site through the language of its film media. With edits that include the direct playback of materials on 35-mm film editing machines, VHS players, and digital audio-visual capture devices, the film also carries within it the remnants of technologies that may one day be vital points of record for remembering our nuclear actions. The external shots in the film are from Dessel, Belgium and focus on the expansive grounds of Tabloo, a nuclear waste visitors centre, which was constructed as a part of the wider development of a large-scale, above-ground storage facility for the low-level nuclear waste products of the Belgian nuclear energy industry. In describing this practice as it developed throughout my research partnership with the BFI, I have been able to trace a path between these two different 'nuclear archives', linking my radionuclides and the machines and media of the BFI archive, to Dessel's radioactive waste depository. As well as documenting and presenting each of the significant locations and sites of my practice at the BFI, the film explores how the activation of these materials, both within and away from the archive site, can open up new avenues for connecting film and media archives with the challenges presented by our interactions with nuclear materials.

Notes

1 The V-Force aircraft (Valiant, Victor, and Vulcan Bombers) were nuclear capable Royal Air Force (RAF) jets, recognised by their distinctive 'V' shaped wing design. See Jones (2000).

2 In my thesis, I argue for an expanded definition of atomicity or atomic-weight that can be equated to the many unfolding impacts of atomic events. See Beck (2024).

3 See https://www.fringeartsbath.co.uk/radiant

4 Peter Cusack's project of 'sonic journalism' is one example of artists working within sites of nuclear significance. See Cusack (2012).

5 See https://curium-244.iless.eu/

6 See https://krypton-85.iless.eu/

7 See https://plutonium-238.iless.eu/

8 See https://plutonium-240.iless.eu/

9 See https://plutonium-242.iless.eu/

10 See https://strontium-90.iless.eu/

11 See https://caesium-134.iless.eu

12 See https://caesium-137.iless.eu/

13 *The Nuclear Archive: Sound Map* is available to download at https://www.iless.eu/Nuclear_Archive-Sound_Map.pdf

14 Steenbeck are a German manufacturer of 35- and 16-mm flatbed film editing tables.

15 A sonograph image visualises the intensity of frequencies within an audio signal over time. The horizontal axis represents the length of the audio signal and the vertical axis represents the frequency in Hz, starting at the bottom with low frequencies and high frequencies at the top.

References

Beck, Daniel Robert. 2024. *Post-Atomic Auralities: A Practice-Based Study of Sound and Atmospheric Effects*. London: University of the Arts London.

Cusack, Peter. 2012. *Sounds From Dangerous Places*. Thornton Heath: ReR Megacorp.

Devine, Kyle. 2019. *Decomposed: The Political Ecology of Music*. London: MIT Press.

Jones, Barry. 2000. *V-Bombers: Valiant, Vulcan and Victor*. Ramsbury: Crowood Press.

Kuchinskaya, Olga. 2023. *The Politics of Invisibility: Public Knowledge about Radiation Health Effects after Chernobyl*. London: MIT Press.

Schuppli, Susan. 2016. 'Trace Evidence: A Nuclear Trilogy'. In: E. Carpenter, ed., *The Nuclear Culture Source Book*. London: Black Dog Publishing, pp. 37–42.

Schuppli, Susan. 2025. Trace Evidence – Susan Schuppli. Available from: https://susanschuppli.com/TRACE-EVIDENCE [Accessed 4/25].

4 A Wylfa Souvenir, Reactivating Nuclear Presence

Helen Grove-White, Bridget Kennedy, and Robin Tarbet

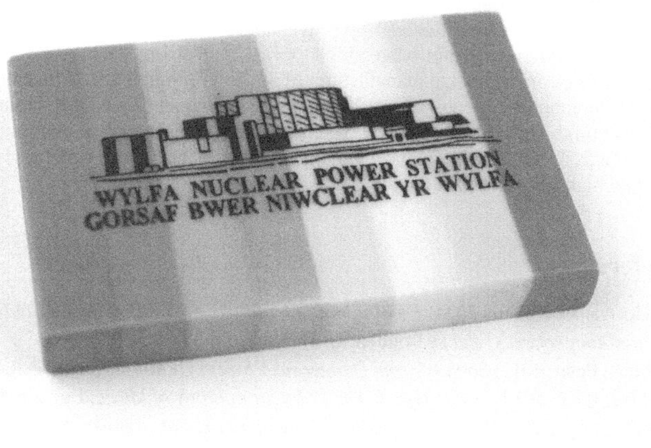

Figure 4.1 A multicoloured souvenir rubber (eraser) featuring the Wylfa Nuclear Power station.

In this collaborative text, three artists come together to contemplate the indexical nature of the Wylfa rainbow rubber (see Figure 4.1) in relation to their shared experience of the *Power in The Land* project[1] (2014–2018). In early 2014, ten artists, selected for their interest in making creative responses to the last active fission generator in Wales, undertook extended enquiries into the temporal and material consequences of this period of transition. As Wylfa A approached shutdown and decommissioning, planning and groundwork for Wylfa Newydd was taking place, generating a productive atmosphere for creative encounters that resulted in a touring exhibition, publication, and an exhibition of archival materials.

DOI: 10.4324/9781003622185-5

A decade later, this rainbow-striped, pocket-sized souvenir provides a springboard for a new set of discussions around nuclear energy and what remains of the Wylfa nuclear power station. Circulating the silhouette of a nuclear power station on a gift shop souvenir runs counter to the fear and anxiety induced by images commonly associated with nuclear physics, such as the mushroom cloud. The familiar and prosaic characteristics of this object as an item of novelty stationary offers 'a measurement for the normal' (Stewart 1993, p. 134) to nuclear technology, and its mass-produced status enables this association to exist as a shared experience. Three of the artists, Helen Grove-White, Bridget Kennedy, and Robin Tarbet, here reflect upon their continued fascination with 'nuclearity' (Hecht 2007, p. 100), a quality that is distributed amongst things, people, and places associated with the uranium fuel cycle but that are not necessarily radioactive. This essay will also explore how their individual creative practices draw on nuclear landscapes and their associated artefacts. Helen, Bridget, and Robin each work in a range of media, with photographic, sculptural, and print processes playing a key role in their respective practices. Through a series of curated conversations, taking place online over a period of 2 months, the artists explore how this small object might represent a powerful industry, as well as the potential of the souvenir to disseminate constructed meanings into the everyday domestic sphere.

* * * *

Bridget: Robin, can you recall how you first came across this object, why you were drawn to it, and why you chose to include it in the *Radiant Objects* exhibition at Fringe Arts Bath in 2023?

Robin: Maybe as a momentary act of being encouraged to buy a novelty gift to mark a nice day out, or subconsciously something more artistically contrived – I was drawn to purchasing a souvenir rainbow rubber stamped with the silhouette of a nuclear power station. I liked the absurd contradiction of the radiantly colourful rubber which seemed in complete contrast to the hostile experience we had walking around the periphery of the site. It was my first visit to a nuclear power station and seeing the scale of it opened up a subject so immense, I found its timescale and toxicity far beyond my comprehension. But I was intrigued [...].

I do not see the Wylfa rainbow rubber as an 'art' object in itself, more an object of nuclear culture because of its association, but I like the contextual implications of using it as a real artefact to accompany my own art works created from Wylfa. I included it in the exhibition *Radiant Objects* in place of my film titled 'Orange Buckets' that previously accompanied the concrete sculptures I made by casting segments of the Wylfa landscape. As a genuine souvenir purchased from the Wylfa visitor centre it introduced a hint of historical grounding, visually describing the location where the artworks were made and allowed the audience further understanding of the source of the installation.

At the time of purchasing, it was a souvenir from a souvenir shop. But now 10 years on, since the visitors centre has been demolished and the power station is being decommissioned, the rubber feels more like a surviving artefact from a physical site that is being slowly dismantled or mothballed. As an object, the rubber is a relic that continues to provoke my wider thinking into nuclear issues.

Helen: I want to add to the subject of nuclear tourism. During the first phase of building Wylfa, a viewing tower was erected for visitors to see the spectacular work of construction, presented to the public as an exciting and wholly positive development. This was then followed, in the early 80s, by a viewing platform on the wooded mound, very close to the reactor building and accessed through the public nature trail. Nature became part of the visitor offer very early on. Subsequently a visitor centre, hosting schools' educational programmes, interactive science displays, a cafe and a gift shop, was set up. Is it this hinterland, the nuclear 'good news', that lends potency to our multi-coloured rubber, the appealing graspable souvenir? I wonder whether the rubbers have been part of the propaganda, symbolically radiating and dispersing the good vibes, entering homes and children's lives through pocket money pick-ups and reflecting that desire to come away with something of the excitement of the visit to the unfamiliar and otherwise ungraspable.

Robin: From all the public outreach you mention, Helen, combined with the souvenir memorabilia, there is no doubt that Magnox Ltd (the company who managed Wylfa) had aspirations for their nuclear power station to be a proud site of modern industry, celebrating the dawn of clean energy amongst nature – a fun day out for all the family with a gift shop. In this way it is perhaps less likely to be seen as a threat. Helen, you and your family have lived close to the power station since it was built in the 60s. What was it that prompted you to initiate this multi-artist exploration around the time of the decommissioning?

Helen: The arrival of Wylfa in the 1960s brought dramatic changes to the area, both physical and cultural. By the time I came here in 2005 I was firmly of an anti-nuclear persuasion and chose, at first, to block it out and ignore it. I was, however, concerned about nuclear waste, nuclear weapons and about the wider implications of the industry, including security and surveillance. Once I stopped ignoring these concerns, I had to find ways to address them as an artist. I found them too overwhelming to address alone, hence putting out a call to artists for a more collaborative approach. My prime concerns were then, as decommissioning approached, about the scale and time-scales of the residue and the legacy of the industry at the end of its life – I still like to think of it as the last nuclear power station in Wales, Wales having declared itself nuclear free in 1982.

Each of the nine other artists brought a fresh and different approach to the subject and on each field trip and weekend gathering there was shared dialogue and new perspectives. What had been fairly familiar to me was clearly not to others, particularly to those in generations younger than myself who had not grown up with Cold War fears and a background in environmental awareness. Paradoxically, I had myself overlooked the visitor centre with its educational and propaganda perspective and had not considered the creative possibilities afforded by the souvenir or indeed of the humour and irony that later featured in my work. I had been focussed on nuclear materiality, whereas the immaterial quickly became the focus for some of us, highlighting questions of visibility and invisibility. In other words, I was initially thinking of the 'big stuff'; scale, power, toxicity and timescales initially and not tuned in to the public interface, even less so to the small and seemingly mundane.

Robin, on that first occasion we went into the gift shop, were you looking for something more accessible? Tell me what appealed there and how it fed into your own work, which in the end involved the material rubber.

Robin: The visitor centre gave me a way in – a way to approach a subject that politically I was sure I was against, but in reality, I knew very little about. And maybe I ended up making the brightly coloured moulds on site because, as with the rainbow rubber, there is an arresting disjuncture between these materials and the weight and seriousness of the nuclear plant.

Our initial nuclear field trip to Wylfa felt like a group of artists visiting an industrial site. We all voyeuristically studied the exterior shells of the weathered complex of anonymous buildings through the very large fence. I did not know that bit of coastline before the power station was built and it seemed quite sensitively settled in, much helped by the colour scheme and landscaping.

In terms of the 'big stuff' such as power, that you mentioned Helen, I was impacted by the way in which the security staff dealt with us – I found their obvious and expected presence often took my focus away from thinking about the core concern of nuclear power generation itself (that was hidden out of sight) and instead drew my attention to surveillance and control. After being accosted by the Civil Nuclear Constabulary (CNC) just for walking down a public road at the periphery of the site, I was not then expecting to 'exit via the gift shop' and be bombarded by footage of a former *Dr Who* actor presenting a promotional video. I found the whole place intriguing, and I wanted to find a way in to further explore it, beyond the visitors' experience, but to do this I knew I would need permission. The souvenir rubber was the trigger that got me thinking about the type of alternative sculptural souvenirs I would make if I could manage to get into the site.

So I wrote letters. I contacted every related email address I could find. The perception of the company running the site was that it was just a normal workplace, so I applied that to my approach. I framed my methodology as an artist in the wider context of materiality and landscape rather than nuclear investigation. I was very forthright about the practicalities of what I wanted to do (making direct casts on the site), and described how I would operate in unthreatening terms familiar to engineers and trade contractors visiting the site. Paperwork was key to gaining access, and it was all about risk assessments and a clear schedule of working procedures. When I was finally invited to a portacabin for an initial induction with other potential contractors, I realised what I was describing seemed the least intrusive behaviour in both procedure and security when compared to the likes of drillers, riggers, archaeologists and environmental surveyors. The main negotiation for me was regarding the containment of my moulding materials when in progress, making anything I temporarily bring onto the site visible and clearly marked, and secure from any environmental contamination. I also had to avoid any of my moulds being identified by the CNC as a suspicious object and mis-interpreted as an attempted terrorist attack, potentially triggering an immediate evacuation.

Bridget: Helen, like Robin you chose to engage physically with the site but in your case, you worked in a way that became somewhat subversive. As someone who lives within walking distance from Wylfa, did you feel able to exploit your position as an insider/local?

Helen: The insider/outsider question falls away when you encounter the reality, as we all did, of the security systems at Wylfa – we were all outsiders. Do you remember before we were given a carefully orchestrated inside visit, we had to submit elaborate details for security vetting (including, for good measure, our foot sizes)? I had already by then encountered the CNC patrol and received the first examination and rather intrusive examination of my video camera. When we went through radiation detection portals and entry procedures, I think the security issues were very real and at the same time, as a local, I had no wish to alienate the no doubt decent local people who were going about their work. I sometimes felt as if I were on a tightrope between my anti-nuclear sentiment and the confronted experience of an establishment clearly out to prove that no safety stones were unturned. I found myself pressing up against the restrictions, literally, by using cameraless photography to make an architectural blueprint of the physical presence of the fence – more 'real' than a photograph. That this was a harmless activity and could be done with the full knowledge of the security staff was part of the irony. So too was the choice of the cyanotype photographic process, the same chemistry used for cleaning up Caesium 137 after the Chernobyl disaster (Brewer 1993).[2] Our choice as a group was not to be provocatively anti-nuclear but many of us played with those edges and chose humour and allusion to draw out our questions and concerns – subversive perhaps, in the eye of the beholder.

Bridget, this leads me on to ask about your research, which continues to focus on issues of nuclear waste. Early on, in our research as visual artists, we had to confront both the visible legacy of the plant and the invisibility of both electricity and radioactivity. Your work at the time involved both containment and transmission, bentonite and copper, as I recall, material approaches to the invisible.

Bridget: The invisible nature of radiation is one of the factors that makes it such a menacing force in the social imaginary, so much so that it has become a regular 'baddie' in Hollywood action movies. However, this is also why it is so fascinating to me as an artist. Both radiation and electricity are kind of magical to me. Although I understand the physics and chemistry involved, I am not a scientist, I want to remain in a magical space of not knowing, it is this ambiguity that attracts me.

The materials I chose to explore in *Power in The Land* were related to the containment of radioactive wastes, so I was thinking about the deep time future of the fuel rods that would be coming out of Wylfa during the decommissioning process. Copper and bentonite clay will be used in deep geological repositories that are designed to keep radioactive waste away from living things for thousands of years. The presence of these materials in the ground could be thought of as making radioactivity visible, but you would have to know what their function was and why they were there in the first place, which is unlikely if you had already dug down into the ground and discovered them, unless you had plans to recover the waste.

I think there are different degrees of visibility in relation to radioactivity. Depending on your perception of the institutions that manage radioactive products, your knowledge of engineering and science involved, and your first-hand experience of nuclear landscapes, places like Wylfa may be in sharp focus or a hazy, fuzzy blur of fear, misunderstanding or disinterest. To many, these nuclear places appear exceptional, full of ominous power and malingering toxicity. To others, everyday experience renders them less visible, or less intimidating at least. This is why I think the rainbow rubber is such a potent nuclear object in the way that the cheerful colours make it highly visible, but through its function as an eraser it is inextricably linked to invisibility.

Robin: Bridget, do you now see decommissioned Wylfa as a physical facade that masks the vastness of deeper hidden concerns of nuclear? You went on to explore Sellafield in great depth where waste from Wylfa is physically stacked up awaiting the promised solution. Is the tiny rubber a proxy unit for more sinister remains?

Bridget: As a nuclear power station is decommissioned it feels like the end, maybe, especially to local people who see the site and its associated jobs come and go. However, I cannot think of Wylfa without thinking about the potential of another power station being built there in the future. Although plans for Wylfa Newydd are on hold it has always been a shadowy presence in my mind.

I think that all nuclear energy projects hide behind the façade of carbon free generation, and whatever the future holds in terms of nuclear technology, the waste that exists now, the legacy of over half a decade of nuclear energy, is still a real, material, unresolved issue. With energy crises looming even some nuclear sceptics have been changing their tunes, and despite the huge costs involved, it is possible that nuclear power stations that were put on hold due to financial issues, such as Wylfa Newydd, might finally go ahead. It is difficult to imagine life without electricity, now actively promoted as a clean, greener alternative for heating your home, for powering your car etc. Beyond the issue of where the legacy waste is going to be stored in the long term, and we are talking thousands of years here (Nuclear Energy Agency 1989)[3], there is a more current issue around mineral resources. For example, Greenlanders have been facing a dilemma over their mineral rich heritage because money from mining uranium and other rare earth metals holds the promise of financial independence and security for the islanders, but at what cost to the environment?

In general, it feels that 'we the public' struggle to comprehend the longevity of the nuclear materials. Once the spent fuel rods leave the power station they are somehow erased from the everyday, they no longer power your toothbrush or your toaster. Stored away, these materials are like a dirty secret that most people gladly forget. The scale of the nuclear waste housed at Sellafield, where the spent fuel rods from Wylfa were sent, is difficult to imagine. But this is our job as artists, we deal in materiality, we search for tangible traces and ask ourselves: How can we make this complex situation comprehensible through visual means? Great British Nuclear (GBN), formerly British Nuclear Fuels Limited (BNFL) proposed a small round disc as a physical representation of the amount of high-level radioactive waste an individual's lifetime use of nuclear energy would produce. At the Wylfa Visitor Centre an information video suggested a suitcase full of intermediate level waste, other sources reference a carrier bag. As for the rainbow rubber as a proxy unit, I am not sure what it might represent but it does work as an 'interscalar vehicle' that has the power to 'open our imaginations' (Hecht 2018, p. 115) to the debate around nuclear energy, keeping it visible and present.

To conclude, our souvenir has entered numerous family homes as a reminder of an exciting day out and an encounter with what was once seen as cutting-edge science. Some of these rubbers may have been put to use, slowly losing their crisp edges, and becoming a bit grubby as they erase past mistakes. Some may be lurking in an old pencil case or at the back of a drawer, long forgotten about. By choosing to explore what Gabrielle Hecht calls 'nuclearity' in the Wylfa rainbow rubber, our encounters with this object have revived wide ranging conversations that were originally inspired by our experiences of the site itself. We see it as a visible reminder of the far reaching social and political implications of the nuclear industry. By exploring what Susan Stewart refers to as the 'double function' of the Wylfa rainbow rubber we have revisited the past and explored how this small object might function to subvert or 'discredit the present' (Stewart 1993, p. 139). Dug out of the drawer, dusted off, and propped up on a shelf or a mantlepiece, it acts like a tiny beacon shining a light on the ongoing questions about the future of the Wylfa site. As our discussions come to an end, for now, we speculate as to whether nuclear energy projects of the future will have visitor centres with gift shops attached to them, and if so, what will they stock?

Notes

1 *Power in The Land* was initiated by Helen Grove White. See https://web.archive.org/web/20220816035743/http://powerintheland.co.uk/

2 Prussian Blue, or potassium ferricyanide, was used experimentally in the aftermath of the Chernobyl disaster in Ukraine, Sweden, and Snowdonia for its properties of chemically locking in caesium 137 and potentially rendering grassland-fed animals safer for consumption.

3 Nuclear waste materials, such as spent fuel rods from reactors, contain many different radioactive substances. These have different rates of decay or half-lives, the time it takes for the alpha, beta, and gamma rays that they emit to reduce to a level that is acceptable from a health and safety perspective.

References

Brewer, Kevin. 1993. 'The blue hills of Snowdonia'. *New Scientist*, 29 May. Available at: https://www.newscientist.com/article/mg13818751-600-the-blue-hills-of-snowdonia/ [Accessed 7/25].

Hecht, Gabrielle. 2018. 'Interscalar vehicles for an African Anthropocene: On Waste, Temporality, and Violence'. *Cultural Anthropology*, Volume 33, Issue 1. Available at: https://doi.org/10.14506/ca33.1.05. [Accessed 7/25].

Hecht, Gabrielle. 2007. 'A Cosmogram for Nuclear Things'. *Isis*, Vol 98, No 1. The History of Science Society. Available at: https://doi.org/10.1086/512834. [Accessed 7/25].

Nuclear Energy Agency. 1989. *An Analysis of Principal Nuclear Issues*, NEA Issue Brief No. 3. January 1989. Available at: https://www.oecd-nea.org/brief/brief-03.html [Accessed 7/25].

Stewart, Susan. 1993. *On Longing, Narratives of the Miniature, the Gigantic, the Souvenir, the Collection*. Durham and London: Duke University Press.

5 Christian Kuhtz' *Einfälle Statt Abfälle, Sonne-Heft 2 (Invention Not Waste, Sun-booklet 2)*

Marc Herbst

I first noticed Christian Kuhtz' *Einfälle Statt Abfälle, Sonne-Heft 2* (*Invention Not Waste, Sun-booklet 2*) (1998) on a shelf by the beets at a locally owned organic food store in Leipzig, Germany (Figure 5.1). It was in a stack of other similar-appearing books – all A5 booklets that could also be described as well-produced zines. They were all from the same distributor, Packpapierverlag (Packing Paper Press). I was eying Kuhtz' book for its hand-drawn cover and also a copy of Pierre Clastre's *Society Against the State* and asked the store-clerk about them; 'Oh yeah, Packpapier has a great catalog of original books and reprints. Every-so-often, they come by in their old van full of books for a sales-trip. It's always exciting'.

Besides being a part of an exciting distribution programme, I was attracted to *Einfälle* because of its manifest connection to the grassroots German anti-nuclear movement. My eyes were open to cultural objects from anti-nuclear movements due to my co-curatorial role in the SALT. CLAY. ROCK. research project. SALT. CLAY. ROCK. was a several-year-long research project ending in an art exhibition at Berlin's nGbK Gallery.[1] The research investigated rural communities impacted by nuclear infrastructure in Hungary and Germany. We had already visited Hungarian nuclear facilities in Paks and storage facilities in Bátaapáti. In terms of opposition, we had focused on the 1980s grassroots resistance to a repository in the small town of Ófalu.[2] Ófalu's sparsely documented protests[3] seem to have more in common with former East German dissident and environmental movements[4] than those of the 1960s New Left that I identify in Kuhtz's work.

Einfälle Statt Abfälle Sonne-Heft 2 is the second book in Kuhtz' *Sonne-Heft* 'sun-booklet' series. *Einfälle Statt Abfälle* is the overall series title, with six subcategories focusing on sun/solar power, wind power, oven building, bikes, craftwork focusing on pottery and natural home construction, and a series on home concerns like compost toilets and muesli. Its values run counter to consumer society and its greenwashed form. 'I submit that the quality of our lives improves considerably if we are willing to take some control of more areas of our lives, and ideally, to do it within the cycle of nature' (Kuhtz 2012, p. 519).

DOI: 10.4324/9781003622185-6

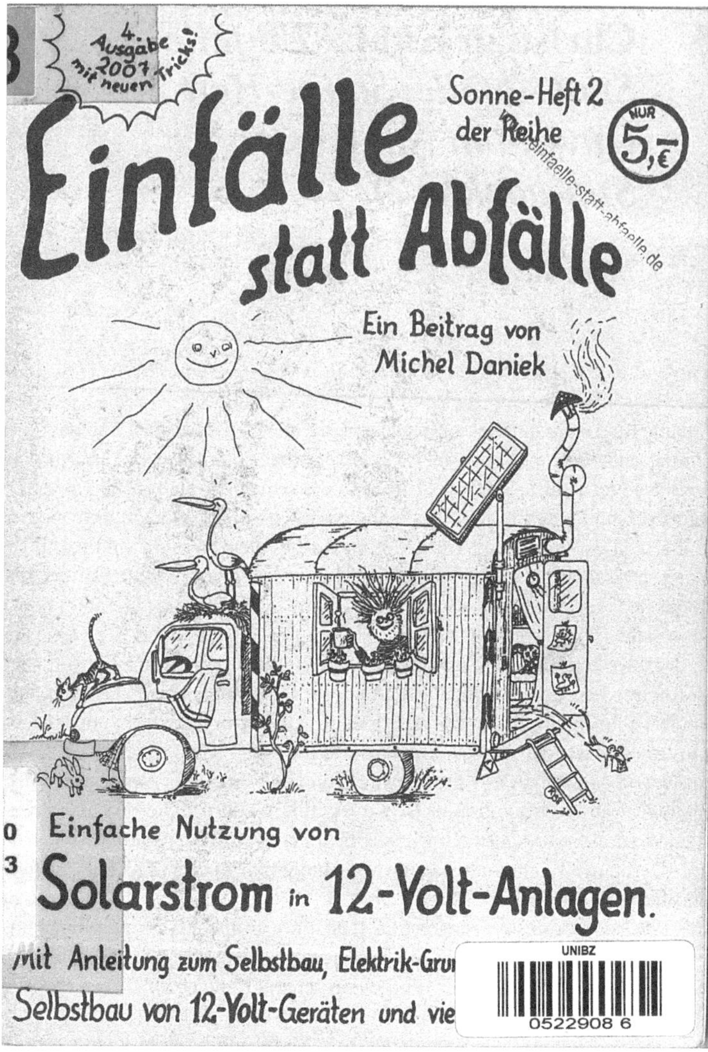

Figure 5.1 Christian Kuhtz' *Einfälle Statt Abfälle, Sonne-Heft 2* (*Invention not Waste, Sun-booklet 2)* © Packpapierverlag.

The hand-drawn cover of *Sonne-Heft 2* features hand-lettering and schematic illustrations that scream 'counter-culture'. There is a certain Robert Crumb-quality in Kuhtz' drawn lines and the pregnant appearance of most of the characters and objects he draws. With the abundance of smiling storks and the cover's twisty chimney, there is also something reminiscent of surrealist-adjacent George Herriman's *Krazy Kat* comic strip. Kuhtz self-distributes the booklets with mail-order info on the inside back cover, and tolerates the proliferation of pirated copies (Kuhtz 2012, p. 519) online in addition to distributing through the above-mentioned Packpapierverlag.[5]

One hand-drawn sun with the shape and smile of Anne Lund's ubiquitous 'Atomkraft, Nein Danke (nuclear power, no thanks)' sun[6] graces the booklet's front cover. There is another on the back cover; it holds a hatchet in one hand, cutting wires leading from a frowning nuclear power plant. While its left hand is busy chopping wires, its right side is beaming down on a solar panel that is hooked up to an electric charge regulator. From there, attached wires go in two directions: one connects to a tangle of overhead lights, projectors, a welding set, circular saws, and other household tools. Below each object are written the interior page numbers where you can learn practical and scientific facts related the objects' usage with solar power. For example, there is a large technical section about a voltage converter, and a short section on a projector – including the hint that broken projector bulbs can be replaced with car headlight lamps (Kuhtz 1998, pp. 49–51, 84).

The wires also lead to car batteries that are used to store gathered and converted solar energy for later use. The car batteries are marked as poison by the skull and bones emblazoned upon them. Kuhtz' series of books reflect a deeply ethical sensibility, reflective of the ecological and economic costs of overconsumption. When discussing the car batteries one should use to store solar energy, he implores the reader to use salvaged batteries, because otherwise their manufacture is just too poisonous.[7] Overall, this and other series books demonstrate a focus on ethical, ecological living within but beyond a consumer-based society. 'As an upstanding citizen, I mind my electricity meter and buy myself a new energy-efficient appliance instead of repairing the old one. But the meter doesn't show me that it often takes far more energy to manufacture the new appliance than I could possibly ever save by replacing the old one …' (Kuhtz 2012, p. 518).

Knowing what became of so much 1960s counterculture[8], one might be tempted to dismiss Kuhtz' degrowth ethic as just being some kind of ethical consumerism. Cynicism around green consumerism is especially valid in the German context that Kuhtz writes for. Germany, the country with the world's third-largest economy, is also the one place that shuttered its nuclear industry due to popular demand after the disaster at Fukushima. The transition away from nuclear power occurred under Angela Merkel's centre-right government that was famed for its stewardship of the German economy through 2008 economic crisis. Merkel is no anti-capitalist. Rather, green politics have been mainstreamed in Germany. The recent strength of the Green Party in electoral politics had them twice acting as a minority partner in the federal government's ruling coalition in the past 20 years. Its most recent governing coalition, from 2021 to 2024, was the so-called 'ample' (traffic light)[9] coalition with the Social Democrats and the Free Democratic Party. That the Greens would form a coalition with the liberally economic Free Democratic Party is evidence enough that the governing realities of Greens in power is not degrowth radicalism or countercultural dreams but rather, green capitalism. Rather than working towards anti-capitalist degrowth economics, the Greens have pushed for green energy transition projects that, at the EU level, are business as usual (Söding 2023), which can be seen as bending towards being anti-labour,[10] and have even been argued as ambivalent to the ecological costs of green transition.[11] While there is nothing to dismiss about the very real phase-out of nuclear power in Germany, this does not mean that the green transition is towards some degrowth utopia.

The following is a small taste of what the clean and green energy transition looks like at the German governance level. To fund our SALT. CLAY. ROCK. research project, we applied for and were awarded a sizable Zero Fund project grant from the German Federal Cultural Foundation that came with the caveat that our project needed to be climate neutral.[12] The granting project was conceived as an experiment to analyze the challenges that cultural producers and institutions face when producing culture in a carbon-neutral way. Our hosting organisation, nGbK, was a smaller institution among the grantees.[13] The funding effort included a 'Zero Academy' intended to train cultural institutions about climate accounting and other issues.[14] One large part of our learning curve was dealing with facts related to how carbon accounting deems cultural projects that spend money on carbon-negative things as more Green than projects that spend money on climate-warming items. The accounting methods have a pro-growth bias as they continue to account for spending rather than evaluating possible cultural impact. Thus, a project that calls for endless industrial growth within a green economy produced in a carbon-neutral way is deemed more carbon-neutral than a small degrowth project that uses normal non-recycled paper within their project.

So, of course, we used an electric car for our trip visiting rural German visiting towns impacted by nuclear infrastructures. When we arrived in the late afternoon at Morsleben, a nuclear waste repository founded by the former East German Government, we were tired and hungry from the day's research visits. Our e-car was low on energy, and we needed to eat. An ironically simple solution was found when we heard that the McDonalds by the edge of town was next to a gas station with a high-speed e-charging stand.

The twin mineshafts in Beensdorf and Morsleben, collectively known as Morsleben, was commissioned as a long-term repository by the Socialist government before the fall of the wall. It was situated in a relatively isolated area near the border with the former West. Unlike many other East German facilities and factories, its staff's salaries were rapidly increased to match those in the West after the fall of the wall. This happened in parallel with the questionable continuation of underground waste depositing, beyond the limitations of its license (Mignon Kirchhof 2018, p. 163–165). Despite these uncertainties underground, Morsleben presented itself to us a stable middle-class community whose residents had managed the rocky political transition well.[15]

On the same SALT. CLAY. ROCK. trip, we also visited the Wendland region. Within the Wendland is the town of Gorleben that is home to a short-term radiation storage facility. The Wendland is relatively close to Morseleben,[16] yet worlds apart. Gorleben was planned in the 1970s to house both a long-term and short-term waste storage facility but due to incredibly strong popular resistance, only the short-term storage facility was built. Transports of all material eventually stored there were met with fierce resistance from the 1996 until the last shipment in 2022, when the last 'Tag X' (Day X) days of Castor transport blockade protests were held. The symbol for these blockade days was an X made of wood[17] which has evolved to become a general symbol for climate movement blockades.

In terms of resistance culture, Gorleben is also notable as the home of the Republik Freies Wendland (Free Republic of Wendland) activist camp that took place in 1980.[18] Over the course of more than a month, the activist camp aimed to block test drilling efforts to find an adequate site for the proposed depository. The camp attracted over one thousand permanent occupiers who built their shelters for the town along with its own radio station, farm, salon, solar power generators, and other social and cultural amenities. It issued, and still issues, passports that are utilised within certain activist communities today.[19] The occupation was eventually cleared out by what was then the largest police operation in Post-War Germany. The image of the Free Republic's lives on and is credited within the German context as the inspiration for the eco-village movement (Andreas and Wagner 2012, p. 135), the movement to build intentional communities that experiment with other ways of living with nature, outside capitalism.[20]

Before it became common practice within Germany to hook up solar power to the national grid,[21] Kuhtz began working on small scale solar for home and community use for the *Sonne-heft* in 1998. Today, he is as skeptical about the value of nation-based carbon accounting[22] as he was then skeptical about the energy industry's ecological motivations for any kind of alternative energy – 'The proven policy of the electricity suppliers to create unlimited demand and then cover it profitably also does not contribute to a more conscious and economical use of energy' (Kuhtz 1998, p. 1). Rather, he noted at the time that solar innovations make most sense for alternative communities.

> Solar collectors for heating systems become financially viable after a few years. Grid-connected solar power systems only pay off in the long term, as in Germany only the now statutory 17 pfennigs per kilowatt hour are

> usually remunerated. This is not the only reason why the majority of solar power systems are so-called stand-alone systems with battery storage. For many people, solar power has already become a matter of course. Most solar power systems can be found in alternative housing projects such as caravan parks, project centers and in shared housing, where they are a way for many people to meet their need for independence and an ecologically responsible way of life.
>
> (Kuhtz 1998, p. 2)

Caravan parks, 'wagonplätze' in German, are a particular kind of trailer park. Its homes are improvised from animal trailers, old trucks and buses, horse stalls and construction site cabins, and made livable throughout the year. Some trailers have indoor plumbing; many rely on common water hook-ups and outhouses. Some have solar cells or windmills. These and the alternative house projects that he mentions are synonymous with Germany's radical left. These places often function as small villages and often host events.[23] They are the downscale and punk cousin of the eco-village, places where people choose inexpensive and creative housing options rather than careers that afford them commercial comforts. Living off the normal grid, often within the city, challenges all kinds of material and social invention; cold winter nights invite residents to either clear the snow off their solar panels or to just cuddle with neighbors for warmth.

It may be easy to dismiss of people living in these non-traditional ways as inconsequential. But I would ask the reader to see Kuhtz, and the individuals and communities that utilise his books, as part of something more than simple lifestyle activists. It is no secret that nuclear power and climate change are popular concerns. Kuhtz talks about being an inventive and curious child, and with the oil crisis of 1972 coming into ecological consciousness around both energy independence and the wastefulness of modern consumer culture, he attended antinuclear protests in the 1970s and wanted to figure out what he could do to build alternative futures, rather than nuclear and capitalist ones, in addition to protest.[24]

It is notable that Kuhtz work is reflected in communities that take these issues seriously and act individually and together to solve problems at home in creative ways that they can manage. Kuhtz' approach shows how social changes, even large-scale changes such as denuclearisation, require not just dissent and responsive legislation, but also ongoing creative practices that pave the way for different cultures that foreground care for materials as well as people. And care for cultures that can keep a movement's critical sensibilities alive even after some of their most manifest issues appear to be solved – even as the wider problems they seek to solve remain largely unaddressed.

We rightfully look to the margins of power to appreciate decolonial practices, and as academics are trained to take seriously the efforts of non-Western peoples to invigorate non-Western cultural habits and practices. Considering the ethical and ecological challenges we globally face, also within Europe and the West, it is worthwhile to recognise the radical efforts of those who are also trying to do things differently and to appreciate the depth of their ethico-political commitments as individuals and communities as well.

Kuhtz and the practitioners who read his books continue to experiment with other ways of being and organising relationships between the people and the more-than-human world reflected in socio-technological infrastructures that insist on being an alternative to green capitalism. *Einfälle Statt Abfälle* reflects this heritage. It represents a practice that is not the New Green Deal growth-by-any-other-name of mainstream environmentalism.

Notes

1 The curatorial research project that ended in an exhibition, titled SALT. CLAY. ROCK., was funded from 2023 to 2025 by the German Cultural Fund, and awarded a 2-year curatorial research project grant by nGbK (neue Geselschaft fur bildende Kunst) from 2023 to 2024. I was the co-researcher and the co-curator with Katalin Erdödi, Julia Kurz, Virág Major-Kremer, and Vincent Schier. For more information on the project, see https://saltclayrock.ngbk.de/en/about.

2 For information on the Ófalu protest see SALT. CLAY. ROCK.'s research page on the project, https://saltclayrock.ngbk.de/en/blog/research-site-ofalu.

3 Ófalu's former mayor Erzsébet Bechli provided the SALT. CLAY. ROCK. with photos from the protests that she participated in. She is currently trying to publish these photos along with an account of the town's resistance. For more info, see https://saltclayrock.ngbk.de/en/blog/exhibition-research-display-object-14-ofalu-protests.

4 The former East Germany and Ófalu's protest aesthetics seem to mostly utilize black hand-painted with black lettering.

5 Packpapier's website logo has Phineas T. Phreak of underground comic artist Gilbert Shelton's *Fabulous Furry Freak Brothers* sweating at the wheels of printing press, demonstrating the New Left sympathies of the publisher/distributor.

6 For more information on this graphic design object that has come to represent the German anti-nuclear movement, see Nej Tak: https://www.atomkraftnejtak.dk/. For a designer-oriented interview with Lund, one can read https://www.itsnicethat.com/features/anne-lund-the-smiling-sun-graphic-design-internationalwomens-day-08031 (Boddington 2018).

7 Kuhtz comments on the use of recycled car batteries saying, 'please only use salvaged batteries due to highly poisonous manufacturing techniques' (2012, p. 518).

8 When I think of the failures of the 1960s counterculture, I think of Mark Fisher, and the 'Californian Ideology'. Fisher writes about how the collective optimism of the 1960s and 1970s has been turned inside-out into hyper-individualism and collective hopelessness (2019). And in their (1995) *Californian Ideology* essay, Barbrook and Cameron trace how what was understood as 1960s California hippy liberalism transformed into neoliberal techno-determinism.

9 Named for the colours that represent each party that just happen to be those of a stoplight: red for the Social Democrats, yellow for the Free Democrats, and green for the Green Party.

10 See the *Berliner Gazette's* 2023 issue, 'Allied Grounds Toolbox for Exploring the Links Between Labor Struggles and Environmental Struggles' for a pro-labour and pro-environmental look at tensions between labour and the capitalist green transition. For example, see Nedimović (2023) on the complex labour struggle against the privatized "green transition" in Bosnia and Herzegovina that has seen positive developments between labour, ecological, and human rights.

11 Research into the ecological costs for lithium mining and recycling seriously undercut claims regarding the environmental benefits of much of the green economy. See, for example, Dunlap et al. (2024)

12 For more information on the project, see Kultursiftung Des Bundes (n.d.).

13 Cologne's Ludwig Museum and Leipzig's Opera House also participated in this grant.

14 The several structural challenges that we faced as a small institution to assimilate ourselves to Green Accounting methods tailored to large and fee-charging institutions are telling. The Cultural Foundation and its partnered accounting agency were unprepared for collaborations with small institutions that do not charge an entrance fee, as reflected from their calculations that grossly overestimated our carbon footprint. Further, our exhibition was told that we might want to consider having an online-only exhibition in order to meet our climate goals; a suggestion that runs in the face of the ideals for why we have art shows in the first place. For a deeper discussion of the challenges of this cooperation with the Zero Fund of the Germany Federal Cultural Foundation, see https://saltclayrock.ngbk.de/en/co2-logbook.

15 Unless otherwise noted, information about Morsleben is from conversations with long-term Morseleben facility employees and residents.

16 It is rumoured that Gorleben was identified as a storage site for its proximity to Morsleben, in order to spite the East German government. This is often mentioned in conversations about the site, and discussed in Kirchhof (2018).

17 For more information, see Kramer (2018), and also https://saltclayrock.ngbk.de/en/blog/third-article.

18 For two different accounts of the Free Republic, see Elser et al. (2023) and Günter Zint (1980).

19 An informant showed me their Wendland passport that was stamped after an activist engagement related to but not within the Wendland region.

20 Ecovillage scholar Marcus Andreas describes ecovillages as a model village that "consists of the comprehensive attempt to integrate all spheres of life (home life, work, provision, free time) as part of an ecological circular economy (Andreas and Wagner 2012, p. 136).

21 Since the early 2000s, the German Federal Government has supported domestic solar panels and their connection to the national power grid with various schemes. Policies have shifted in relationship to which parties are in office, current events, and current opinions on green energy and the stability of the power grid. The recent "Ample" coalition that involved the Green Party found success with their Balcony Solar project. The Government subsidized easy-to-install solar panels that could be quickly mounted onto the railings of a balcony and plugged directly into a wall so that the energy the panels generated could easily contribute to the national power grid. This project seems to be continuing under the new (2025) government led by the Centre-Right conservatives.

22 Kuhtz suggests that Germany can have lower climate emissions because it offshores its dirty work. Kuhtz (2012).

23 For more on these trailer parks, see Canham (2006).

24 Kuhtz discusses his participation in 1970s anti-nuclear protests in the town of Brokdorf, and the development of his life's work as a designer and author at: https://www.youtube.com/watch?v=9l0GPM84VXI (Arnold 2024).

References

Andreas, Marcus and Wagner, Felix. 2012. '"For Whom? For the Future!" Ecovillage Sieben Linden as a Model and Research Project', *RCC Perspectives*, No. 8, Realizing Utopia: Ecovillage Endeavors and Academic Approaches. pp. 135–148.

Arnold, Gabor. 2023. *Ein Kurzfilm - Nachhaltigkeit im Alltag und im Beruf.* Available from: https://www.youtube.com/watch?v=9l0GPM84VXI [Accessed 07/25].

Atomkraft? Nej Tak. n.d. *The history of the Smiling Sun*. Available at: https://www.atomkraftnejtak.dk/smilingsun/ [Accessed 06/25].

Barbrook, Richard and Cameron, Andy, 'The Californian Ideology', *Mute*, Vol. 1, No. 3. Available at: https://www.metamute.org/editorial/articles/californian-ideology [Accessed 06/25]. Mute Publishing.

Berliner Gazette. 2023. 'Allied Grounds. Toolbox for Exploring the Links Between Labor Struggles and Environmental Struggles'. Available at: https://berlinergazette.de/projects/allied-grounds/ [Accessed 06/25].

Boddington, Ruby. 2018. '"I'm not a designer – I was just an activist": how The Smiling Sun became one of history's most iconic logos'. *It's Nice That.* 18th March 2018. Available at: https://www.itsnicethat.com/features/anne-lund-the-smiling-sun-graphic-design-internationalwomensday-080318 [Accessed 06/25].

Canham, Stefan. 2006. *Bauwagen. Mobile Squatters*. Berlin: Peperoni Books.

Dunlap, Alexander et al. 2024. 'The political ecologies of "green extractivism(s): an introduction'. *Journal of Political Ecology*, vol. 31, pp. 436–463. The University of Arizona Libraries.

Elser, Oliver et al. (eds). 2023. *Protest Architecture: Barricades, Camps, Spatial Tactics 1830–2023*. Zurich: Park Books.

Fisher, Mark. (unpublished) 2019. *Acid Communism (Unfinished Introduction)* Available at: https://my-blackout.com/2019/04/25/mark-fisher-acid-communism-unfinished-introduction/ [Accessed 06/25].

Günter Zint. 1980. *Republik Freies Wendland. Eine Dokumentation, Frankfurt am Main: Zweitausendeins.*

Kirchhof, Astrid. 2018. 'East-West German Transborder Entanglements Through the Nuclear Waste Sites in Gorleben and Morseleben'. *Journal for the History of Environment and Society*, vol. 3, pp. 145–178. Brepols Publishers.

Kramer, Sven. 2018. *Ästhetiken des Widerstands: Literatur und Sprache in politischen Prozessendes deutschsprachigen und arabischen Raums*. Springe DE, Klampen Verlag.

Kuhtz, Christian. 1998. *Einfälle statt Abfälle, Sonne-Heft 2*. Keil Germany: Einfälle state Abfälle Verlag, 1998.

Kuhtz, Christian. 2012. 'No. 081' *Documenta (13) The Book of Books*, catalog 1/3. Ostfildern DE: Hatje Cantz Verlag.

Kultursiftung Des Bundes. [n.d.]. *Zero: Climate-neutral artistic and cultural projects*. Available from: https://www.kulturstiftung-des-bundes.de/en/programmes_projects/sustainability_and_future/detail_1/programm_zero.html#c202239 [Accessed 06/25].

Nedimović, Svjetlana. 2023. 'Fires of the Future in the Furnaces of the Past: Building Socio-Environmental Alliances in the European South(East)', *Berliner Gazette*, May 3rd, 2023. Available at https://berlinergazette.de/building-socio-environmental-alliances-in-bosnia-and-herzegovina/ [Accessed 06/25].

SALT. CLAY. ROCK. Available at: https://saltclayrock.ngbk.de/en [Accessed 06/25].

Söding, Tatjana. 2023. 'Against the End of History: Why We Should Resist the Remaking of Fossil Capital into Green Capital'. *Berliner Gazette*, April 20th 2023. Available at: https://berlinergazette.de/resistance-against-the-transformation-of-fossil-capital-into-green-capital/ [Accessed 06/25].

6 Bunk'Art

Staging the Past and Present of Albania's Nuclear Bunkers

Dimitra Gkitsa

The Bunker as (Unwanted) Cultural Heritage

On a recent trip to Prague, during a tour focused on communist history, the tour group visited a Cold War–era nuclear bunker. As we moved deeper into the underground structure, the temperature dropping and the damp and metallic smell of the air becoming more pronounced, it became apparent that the space was filled with random assorted objects from the communist period: mid-century chairs, rotary telephones, propaganda leaflets and newspapers, a broken bust of Stalin, and an unsettling number of oxygen masks. None of these objects had any captions or explanations as to why and how they had arrived there. What was more bizarre though, in this surreal setting, was the narrative presented by the tour guide. This was not a structured historical account, but a kind of personal monologue. He recounted his memories of life under communism with vocabulary indicating the misery, fear, and distrust that characterised the period. These stories culminated in an account of life in the bunker in the event of an atomic bomb, presented as if it was an *actual event* that took place and, at the same time, as a fantasy that was unfolding without addressing the wider geopolitical context of nuclear deterrence, Cold War militarism, or the global aftermath of Hiroshima. The absurdity was intensified even more with theatrical installations using mannequins attempted to simulate daily life underground (Figure 6.1), reinforcing stereotypes that many Western tourists already carry about life under communism.

This treatment of communism as a personal memory marked by trauma, rather than as a historically and politically complex reality of the recent past, is not unusual. In much of post-socialist Central and Eastern Europe, there is a tendency to frame the socialist past through emotional recollection, moral condemnation, or nostalgic kitsch rather than critical historical engagement. This sentiment has also influenced atomic heritage sites in the region, which more recently have been approached as infrastructures that "endure through their maintenance, through ongoing relations with the promises of past, through a process of valuing, holding on to and adapting what came before against the becoming of the future" (Dawney 2021, p. 417). Nuclear cultural heritage, as

DOI: 10.4324/9781003622185-7

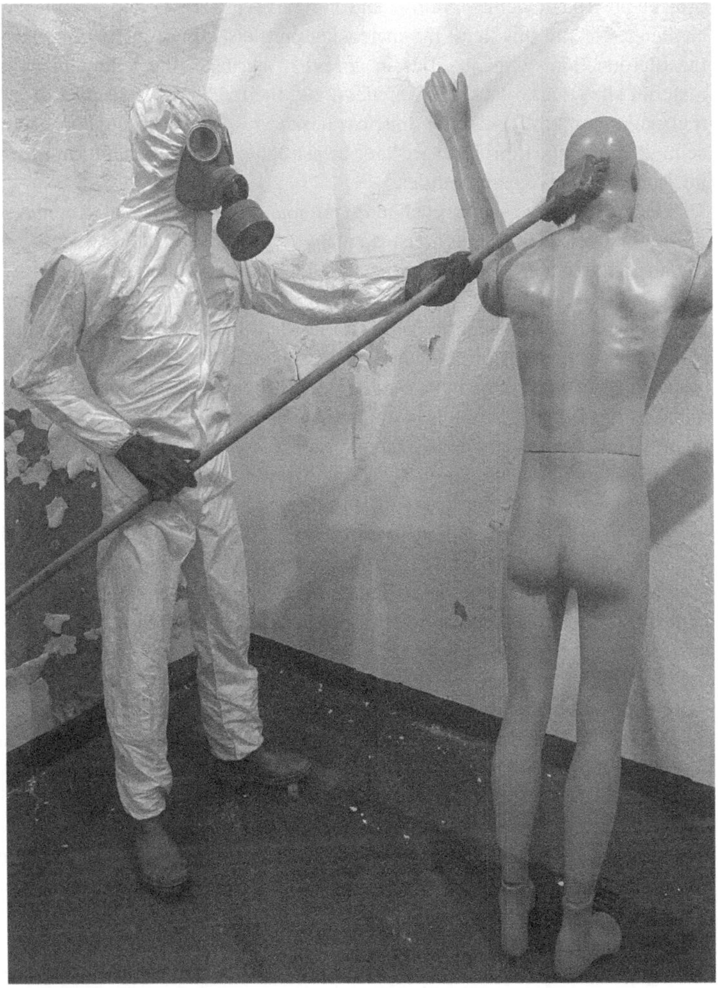

Figure 6.1 Installation view of a nuclear bunker in Prague.

Source: Photo by the author, 2025.

Rindzevičiūtė (2019) describes, is a "vast and hybrid field" that encompasses "complex histories: scientific and technological, industrial, economic, environmental, political, cultural, social and military." These entangled histories make nuclear heritage uniquely charged. The "Nuclear Spaces: Communities, Materialities and Locations of Nuclear Cultural Heritage" (NuSPACES) project by Rindzevičiūtė, Storm, and Dovydaitytė (2024) outlines the key challenges facing this field: the increasing politicisation of nuclear memory, the reproduction of inequalities in access to heritage, the vulnerability of material sites due to lack of awareness, and the tensions between local and transnational narratives. Often, nuclear heritage is framed through Western-centred narratives of progress, risk, or catastrophe, marginalising local meanings and alternative temporalities.

These dynamics become even more pronounced in post-socialist contexts, where the nuclear or military past is entangled with the ideological legacy of socialism itself. Here, nuclear heritage is not passively inherited but actively contested (Ross, 2024). What is preserved, forgotten, or destroyed becomes part of a struggle over identity and historical legitimacy. In such settings, socialist infrastructure is often rendered a dissonant heritage, an unwanted past associated with an illegitimate or discredited regime. As Balockaite (2012) notes in the Lithuanian context, the material remains of socialism are frequently erased or reframed to align with European integration, market liberalism, and national rebirth. The concept of "dissonant heritage," first introduced by Tunbridge and Ashworth (1996), captures this friction. They argue that heritage is rarely consensual; instead, it reflects competing claims over which histories are remembered, and which are suppressed. When multiple groups ascribe divergent meanings to a single site, object, or narrative, heritage becomes contested, unstable, and politically volatile. In this sense, the socialist past becomes dissonant and unwanted heritage because it sits uneasily within the construction of a new ideology imposed after 1989, which seeks to align more with European, liberal-democratic, or capitalist narratives.

Within this broader post-socialist field, Albania represents a particular case. Despite its extreme Cold War isolationism in the 1980s, Albania mirrored many Eastern Bloc practices of militarisation and defence as well as technology and science.[1] While the country never developed nuclear weapons or reactors, it did produce military infrastructure shaped by the imagined threat of nuclear war. During the 1970s and 1980s, the Communist Party commissioned the construction of over 170,000 bunkers across the national territory, ranging from single-person shelters to multi-storey subterranean complexes and adjusted to specific environmental conditions as well as threats (Kristo, Dobjani, and Shtylla 2024).

After the collapse of the socialist regime in 1991, the bunkers were initially vandalised or ignored, regarded as symbols of paranoia, irrationality, and authoritarian excess. Their scattered presence across the landscape, on beaches, in fields, in city centres, made continuously and materially present a history which the rapid transition to neoliberal democracy wanted to erase from the public space. Yet in recent years, two of the largest nuclear bunkers in Tirana have been transformed into museums: Bunk'Art 1 and Bunk'Art 2. Repurposed as museum-gallery hybrids, these two sites present an immersive experience of the socialist period, structured around the themes of surveillance, political violence, and trauma. The curatorial approach aligns with a broader tendency in Albanian public discourse: to either forget the socialist past or represent it through trauma and authoritarianism. The emphasis on victimhood and fear reinforces a linear narrative of democratic progress, while leaving little room for more ambivalent or complex readings of the period.

Paul Virilio's writing on war infrastructure offers a compelling frame through which to understand the afterlife of the postwar bunker. For Virilio, bunkers are not merely obsolete military structures but spatial condensations of power, control, and fear. They are both protective shells and instruments of enclosure. As he writes:

> The poetry of the bunker is in its still being a shield for its users, in the end as outdated as an infant's rebuilt armor, an empty shell, an emotionally moving phantom of an old-fashioned duel where the adversaries could still look each other in the eye through the narrow slits of their helmets.
>
> (1994, p. 46)

This image captures the bunker as an affective object: obsolete yet resonant, hardened yet vulnerable. Paradoxically, bunkers that were specifically designed against nuclear attack were never used or inhabited. As such, their presence reflects both the potential futures of their past, but also the anxieties of the present reality.

The chapter approaches these bunkers as symbolic objects through the case of the Bunk'Art projects in Tirana. While officially presented as heritage initiatives, these sites are deeply ambiguous. They function simultaneously as tourist attractions, curated archives, and historical simulations. Rather than accepting Bunk'Art's self-definition as a museum, this chapter reads the site across multiple registers: as a form of dark tourism, as a selective archive of violence, and as a cultural fiction that constructs the past for contemporary consumption. Through this analysis, the chapter interrogates how the architectural remnants of socialist defence have been mobilised in the production of post-socialist identity, and how the memory of a nuclear future, once imagined as imminent and totalising, has been domesticated, aestheticised, re-politicised, and commodified in the present.

The Bunker as a Site of Dark Tourism

Bunk'Art 1, situated on the outskirts of Tirana near Dajti Mountain, opened in 2014. Occupying a five-storey underground complex with 100 rooms and narrow corridors, it was originally built between mid-1970s and early-1980s to shelter political and military leadership in the event of a nuclear attack. The interior of the bunker has been adapted into a sequence of exhibition spaces, some of which display historical materials related to the development of the Albanian army and its ideological education and artistic installations, while others recreate the bunker's original functions, including meeting rooms, sleeping quarters, and command centres. The curatorial approach here relies heavily on theatrical and atmospheric devices: dim lighting, sound effects, recreated environments, and staged artefacts evoke a sense of unease, isolation, and claustrophobia. Rather than offering open-ended reflection or historical interpretation, the exhibition is structured to move the visitor through a pre-scripted emotional arc that begins with the militarisation of Albanian society and culminates in the violence of state repression. Yet this progression is curated through affective experience, not contextual understanding. The violence of the regime is presented not through sustained analysis of structures of power or political transformation, but rather through its emotional effects: fear, isolation, loss. In this sense, not only have meanings already been predefined for the visitors, but also the very violence that the site supposedly wants to remember is staged through immersion rather than historical explanation, rendering it a touristic spectacle.

Bunk'Art 2, which opened in 2016 in the city centre behind the Ministry of Interior, follows a similar narrative. This bunker is smaller in scale, comprising 24 rooms, and was constructed as a shelter for the secret police during the socialist regime (*Sigurimi*). Its exhibitions are more focused on state surveillance and political repression, and include reconstructed interrogation rooms. Visitors encounter black-painted corridors and testimonial installations featuring the voices and images of former political prisoners or their relatives. While these installations invoke suffering, they are rarely situated within the broader history of the socialist state, including its legal framework, ideological claims, or international alliances.

This immersive, theatrical mode of musealisation positions the two sites of Bunk'Art firmly within the framework of dark tourism. Often defined as tourism to sites associated with death, violence, suffering, or atrocity, dark tourism involves a consumption of trauma that is structured through emotional, spatial, and symbolic cues. As Lennon and Foley (2000) have argued in their foundational work on the subject, dark tourism is characterised by "the presentation and consumption (by visitors) of real and commodified death and disaster sites" (p. 3). In this framework, historical authenticity is frequently entangled with spectacle. What is sold is not merely knowledge of the past, but the illusion that violence is something that is no longer present in the "new" freedoms of neoliberal democracy. The sites do not simply present the past; they stage it. The visitor here is driven in a carefully orchestrated environment with little room for questioning, re-imagining, or shaping a polyvocal understanding of the past.

The transformation of the bunker into a tourist site mirrors broader shifts in post-socialist Europe, where heritage becomes entangled with processes of marketisation, rebranding, and national repositioning. As Ivanova and Buda (2020) note, heritage tourism in the post-socialist context forms part of the larger transition from state-directed to market-oriented economies.

In this shift, memory becomes a resource used to mediate between past regimes and present aspirations. This transformation is further underpinned by shifting legal and institutional frameworks. As Pompejano (2019) has argued, Albanian legislation on cultural heritage has consistently evolved in step with the political demands of the moment. The definition of what constitutes a "monument" is never static but always reflects the dominant ideology. In the case of Bunk'Art, the former military infrastructure is reclassified as a site of memory only insofar as it can be framed as a space of mourning or condemnation. The Cold War context, its ideological function, and its symbolic complexity are largely sidelined in favour of a narrative that is emotionally legible and politically expedient.

Bunk'Art 1 and Bunk'Art 2, as sites of dark tourism, align with a wider regional pattern in which post-socialist heritage tourism becomes a vehicle for transition – not just from communism to capitalism, but from collective memory to commodified experience. As Isto (2017) argues, the production of memory under neoliberal conditions is not merely about historical recovery; it also performs the "opening of memory-as-tourism to Europe." What is remembered, and how, is shaped by the imperatives of attraction, legibility, and international recognisability. Nowhere is this commodification more starkly revealed than in the museum shop, where even the legacy of repression is available for purchase. Miniature concrete bunkers, scaled-down versions of the very structures meant to shield the elite from nuclear annihilation, are marketed as "souvenirs that hold memories." The irony is piercing: the architecture of fear and violence, once hidden underground, now resurfaces as a novelty object, only this time it is something to be bought, packed in a suitcase, and displayed on a shelf.

The Bunker as a Curated Archive

Albania's abandoned and decaying socialist infrastructures, from bunkers and military infrastructures to ruined factories and memorials, can be read as material witnesses to the many and abrupt transformations that have shaped the country's recent history (Gkitsa 2024). These ruins possess a kind of affective charge that lends them to polyvocal artistic and political engagements. Their very incompleteness, their exposure to reinterpretation, makes them fertile ground for curatorial and artistic intervention. The Bunk'Art project, however, stands apart from these ambiguous ruins. While derelict factories and scattered bunkers offer open-ended spatial encounters and remain subject to conflicting or emergent meanings, Bunk'Art has already been curated into a coherent interpretive frame. Here, meanings have been predefined and structured by a narrative of authoritarian violence, emotional spectacle, and moral clarity. An interesting part of the curatorial strategy is the use of archives and documents, which are juxtaposed and exhibited together with contemporary artworks.

In analysing the exhibition of archival material in highly stylised, darkened rooms with documents treated more as aesthetic artefacts than historical records, Jonida Gashi (2022) argues that this curatorial approach fosters aesthetic appreciation rather than active engagement, staging what she calls "a performance of the separation from (collective) memory and the past" (p. 131). This approach echoes practices seen elsewhere in the region, such as the Museum of Communism in Prague or the House of Terror in Budapest. In both sites, archival materials are similarly embedded within affective environments that prioritise moral clarity and political condemnation. Sara Jones (2011), writing on these museum practices, observes how "ideological imperatives" often replace "individual understanding and meaning-making" (p. 109). The visitor is guided through a curated moral landscape without room for historical nuance or differentiated perspectives. This museological model, while rhetorically powerful, simplifies history to fit a post-socialist narrative of victimhood and liberation.

Notably, Bunk'Art also includes several contemporary artworks exhibited alongside archival documents, historical photographs, and reconstructed spaces. However, these artworks rarely introduce a rupture in the representational logic of the site. Instead, they often adopt similar strategies of visual and emotional saturation. For example, Ledia Konstandini's site-specific installation *Incube* appears as a painted box illuminated by fluorescent lighting in a narrow corridor. The work's claustrophobic enclosure and artificial glow resonate with the bunker's ominous atmosphere, reinforcing, rather than unsettling, the site's affective frame. Similarly, Rajmonda Zajmi Avignon's *Monster of Dictatorship* presents an anthropomorphic figure built from military boots, a television, and a gas mask, the most obvious symbolic fusion of surveillance, media propaganda, and authoritarian control. Installed in one of the dimly lit corridors, it does not invite interpretation so much as it amplifies the surrounding scenography. Such artworks are displayed within sight of archival materials and mannequins in military uniforms, as though they were another layer in the curated script. The integration of art and archive could have provided friction, invited questioning, or opened creative dialogue with history. Instead, as Vincent W. J. van Gerven Oei (2016) critically observes in his review on Bunk'Art 2, the absence of interpretation (perhaps the very outcome of critical research on complicated history which deserves even more attention and care), "makes the communist past so incomprehensible that no one will dare to ask a single critical question – because they wouldn't know where to start." This sense is evoked precisely because Bunk'Art fails to function as a space for learning or unlearning a difficult history, but rather as a staged spectacle in which even trauma is rendered consumable, and memory is flattened into a commodity.

The Bunker between Fact and Fiction

In this context, the curated archive becomes a site of fiction more than a site of historical fact. The boundary between document and installation, between evidence and allegory, is blurred. The visitor experiences history not as a layered, contradictory process, but as a closed narrative of suffering. This undermines the potential of the bunker as a complex site of memory, reducing its historical specificity to a generic experience of authoritarianism and trauma. The Cold War context, the specificities of Albanian socialism, and the ideological tensions of the time are flattened under the weight of an immersive spectacle. As Pierre Nora has previously observed, "there are *lieux de mémoire*, sites of memory, because there are no longer *milieux de mémoire*, real environments of memory" (1989, p. 7). Here, environments of memory include living and organic environments in which memory is embedded, for instance, such oral traditions, everyday practices, communal rituals that are passed down through generations: memory as a lived experience. On the other hand, sites of memory are cultural and symbolic constructions that are created exactly because those living environments no longer exist or function. Sites of memory include monuments, museums, archives and other deliberate attempts to preserve memory after it has ceased to be part of the everyday life.

Bunk'Art illustrates this shift. It is not a continuation of a living memory, but a staging of its absence. The bunker becomes a theatre in which the past is reenacted, often spectacularly, because it no longer lives in everyday collective practice, becoming a sealed environment where meaning is delivered, rather than discovered. Similar sites of memory and dark tourism have been previously viewed as heterotopias; as standing in-between spaces of the everyday and disaster, the real and the imaginary (Stone 2013). Foucault's concept of heterotopias refers to real places that exist in the world but operate as "other" spaces that are outside of ordinary social norms, yet intimately connected to them (1986). Unlike utopias, which are idealised and non-existent spaces, heterotopias are actual sites that juxtapose several layers of meaning, time, and function in one place. As such, they often reflect, invert, or contest the spaces around them. In a similar vein, sites of dark tourism do not simply preserve the past. Instead, and more crucially, through curated artworks or archival displays, they reorganise the past, offering a staged encounter with loss or horror that is shaped by contemporary cultural, political, and even touristic frameworks.

This heterotopic dimension is reinforced by the bunker's own architectural ambiguity. As a Cold War structure built for survival in the event of nuclear attack, the bunker was always imagined in relation to a future catastrophe that never arrived. Its very existence is speculative. It is a space constructed not for everyday life, but for a hypothetical future, which has since been reabsorbed into the past. This speculative temporality is mirrored in its museological form: the visitor moves through spaces that were once oriented towards imagined futures of war and disaster, now curated as haunted remnants of a paranoid past. Fiction and fact fold into one another. In Bunk'Art, this speculative quality is not neutral. It is ideologically framed. The bunker no longer shelters against nuclear annihilation; it now shelters a specific post-socialist memory regime – one that organises the past around a singular moral arc, aligning with contemporary narratives of Europeanisation, liberalism, and capitalist modernity. Its transformation into a dark tourist site does not merely communicate memory. Instead, it manufactures it, using the aesthetics of horror and isolation to shape an emotionally legible, politically expedient fiction.

Perhaps a crucial question that arises here is: for whom do these bunkers exist? The museum's curated experience seems primarily tailored to foreign visitors and the international gaze, evident in the bilingual placards, dramatic scenography, and simplified version of communism. This raises critical concerns about curatorial and museological approaches to sites of nuclear cultural heritage, particularly in the post-socialist geo-political reality. Bunk'Art fails to create a memory of the generation who has lived memories of communism, since their multivocal experiences are never included in the creative or curatorial process. Similarly, Bunk'Art does not provide a collective memory for the generations that are yet to come. In fact, lacking historic information and critical engagement with the past, it renders this complex period even more distant, detached, and abstract. As a result, Bunk'Art does not fulfil the responsibility and ethical labour of memory work. In this sense, it fails not only as a museum, but also as a civic institution tasked with fostering critical engagement with the past. Curating "difficult knowledge," a knowledge that emerges from histories of violence, repression, and injustice, requires more than the display of emotionally charged imagery or top-down efforts by the state. More crucially, such work involves the creation of interpretive spaces in which memory is "mindfully deployed by individuals and groups in attempts to provoke, enable, and transform" (Lehrer, Milton, and Patterson 2011, p. 3). It demands curatorial strategies that resist closure, embrace ambiguity, and invite reflexive, often uncomfortable and unsettling, encounters with the past. Rather than guiding the viewer towards a single moral conclusion, curating difficult knowledge should foster active engagement, allowing for disagreement, critical thought, and the possibility of transformation.

New Possibilities between Present Pasts and Lost Futures

It could be argued that our time is defined by its present pasts. The phrase, offered by Andreas Huyssen (2003), describes a contemporary reality in which we "suffer from a hypertrophy of memory, not history" (p. 3), which is the very symptom of our inability to imagine alternative futures (p. 2). On the one hand, this approach highlights that the past is not fully finished, that it remains active, shaping contemporary identity and politics. On the other hand, it also shows that the ways in which societies remember are deeply influenced by current concerns, anxieties, and power structures. Memory is always mediated by the present.

Bunk'Art emerges from within this condition. It stages the past not as something finished, but as something incessantly present: looped in soundtracks, entombed in corridors, flickering across walls. But this presence is tightly controlled. The bunker no longer protects against nuclear catastrophe; it now shelters a curated narrative in which the imaginary of trauma is more vivid than the histories that produced it. Bunk'Art points to the need to do heritage otherwise, perhaps refusing such closures. It means acknowledging that the bunker, like other Cold War infrastructures, cannot be reduced to a single function or a single narrative. As Iain Robertson (2016) has argued, a "heritage from below" begins not with institutional narratives but with the fragments of meaning that persist in everyday life. In the Albanian context, this might mean listening to how people live amongst bunkers today: how they reuse them, ignore them, dismantle them, or narrate them through stories that escape the museum's frame. It means engaging with the dissonance rather than resolving it. This is not a call to abandon the bunker, but to unseal it symbolically, curatorially, politically. To treat memory not as a fixed object but as a process that is fractured, ultimately polyvocal, and situated. To recognise that the past, especially one shaped by nuclear imaginaries and socialist experimentations, does not end with the state, or with collapse. It continues through adaptation, improvisation, and refusal. The bunker, in this sense, is not only a remnant of a lost future; it may also be a space in which new futures, however precarious or speculative, can still take shape.

Note

1 For instance, as Arjan Shahini (2025) shows in a recently published article that the Institute of Nuclear Physics of the Academy of Sciences, in the 1980s, by being part of the International Atomic Energy Agency, albeit late in comparison to other socialist countries, played a role in introducing cybernetics in the country. Similarly, historians such as Ylber Marku (2023) and Elidor Mëhilli (2017) have challenged the dominant view of socialist Albania as radically isolated from the rest of the world, showing how alliances and communication with the rest of the socialist world was paramount in the development of technology, industry, and science.

References

Balockaite, Rasa. 2012. Coping with the unwanted past in planned socialist towns: Visaginas, Tychy, and Nowa Huta. *SLOVO*, 24(1), pp. 41–57

Dawney, Leila. 2021. The multiple temporalities of infrastructure: Atomic cities and the memory of lost futures. *Environment and Planning D: Society and Space*, 39(3), pp.405–422.

Foucault, Michel. 1986. Of other spaces: Utopias and Heterotopias. *Diacritics*, 16(1), pp. 22–27.

Gashi, Jonida. 2022. Politics of installation as politics of the archive: Three case studies. *Art Studies*, 19, pp. 125–142.

Gkitsa, Dimitra. 2024. Memory matters: Post-socialist wastelands as materials bearing witness. *Art Studies*, 22, pp. 113–155.

Huyssen, Andreas. 2003. *Present Pasts: Urban Palimpsests and the Politics of Memory*. Stanford, CA: Stanford University Press.

Isto, Raino. 2017. An itinerary of the creative imagination: Bunk'Art and the politics of art and tourism in remembering Albania's socialist past. *Cultures of History Forum*. Available at: https://www.cultures-of-history.uni-jena.de/politics/bunkart-and-the-politics-of-art-and-tourism-in-remembering-albanias-socialist-past [Accessed 6/10/2025]

Ivanova, Milka and Buda, Dorina-Maria. 2020. Thinking rhizomatically about communist heritage tourism. *Annals of Tourism Research*, 84, p. 103000.

Jones, Sara. 2011. Staging battlefields: Media, authenticity and politics in the Museum of Communism (Prague), The House of Terror (Budapest) and Gedenkstätte Hohenschönhausen (Berlin). *Journal of War and Culture Studies*, (4)1, pp. 97–111.

Kristo, Saimir, Dobjani, Etleva and Shtylla, Saimir. 2024. Unearthing history: Underground cities and military heritage in Albania and Europe. *Academic Journal of Business, Administration Law and Social Sciences*, 10(3), pp. 1–28.

Lehrer, Erica, Milton, Cynthia E. and Patterson, Monica Eileen (eds.). 2011. *Curating Difficult Knowledge: Violent Pasts in Public Spaces*. New York and Basingstoke: Palgrave Macmillan.

Lennon, John and Foley, Malcolm. 2000. *Dark Tourism: The Attraction of Death and Disaster*. London and New York: Continuum.

Marku, Ylber. 2023. Preparing for an alliance: China's socialist model and Albania's economic path in the Early Cold War. *European Review of History: Revue Européenne d'histoire*, 30(3), pp. 410–432.

Mëhilli, Elidor. 2017. *From Stalin to Mao: Albania and the Socialist World*. Ithaca, NY: Cornell University Press.

Nora, Pierre. 1989. Between Memory and History: Les Lieux de Mémoire. *Representations*, 26, Special Issue: Memory and Counter-Memory, pp. 7–24.

Pompejano, Federica, 2019. From national to cultural monuments: An overview of architectural heritage protection in Albania (1912–1992). *Journal of Architectural Conservation*, 26(1), pp. 55–70.

Rindzevičiūtė, Egle. 2019. *Nuclear Cultural Heritage: Position Statement* (Project Report). Kingston upon Thames: Kingston University.

Rindzevičiūtė, Egle, Storm, Anna and Dovydaitytė, Linara. 2024. *Nuclear Spaces: Communities, Materialities, Locations of Nuclear Cultural Heritage. Concluding Report*. Kingston upon Thames: Kingston University London.

Robertson, Iain J. M. 2016. *Heritage from Below*. Abington: Routledge.

Ross, Linda M. 2024. Nuclear cultural heritage: From energy past to heritage future. *Heritage & Society*, 17(2), pp. 296–315.

Shahini, Arjan. 2025. Cybernetics on the margins: Computer science in Albanian higher education during the Cold War. *Cold War History*, 25(1), pp. 89–104.

Stone, Philip R. 2013. Dark tourism, heterotopias and post-apocalyptic places: The case of Chernobyl. In L. White and E. Frew (eds.), *Dark Tourism and Place Identity*. Melbourne: Routledge, pp. 79–94.

Tunbridge, John E. and Ashworth, Gregory John. 1996. *Dissonant Heritage: The Management of the Past as a Resource in Conflict*. New York: John Wiley and Sons.

Van Gerven Oei and Vincent, W. J. 2016. Bunk'Art 2: A nuclear attack on meaning. *Exit*. Available at: https://exit.al/en/2016/11/21/bunkart-2-a-nuclear-attack-on-meaning/ [Accessed 15/06/2025].

Virilio, Paul. 1994. *Bunker Archaeology*. New York: Princeton Architectural Press.

7 Contaminated Tartan

Michael Sanders and Jonathan Faiers

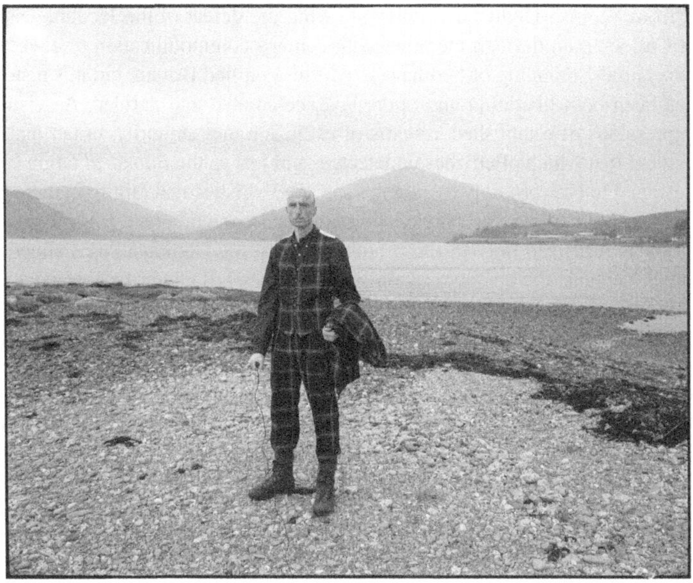

Figure 7.1 Artist Michael Sanders wearing the *Polaris Military* tartan suit he designed, standing across the bay from Coulport, north of Holy Loch, where Polaris missiles were housed and maintained during the Cold War. Image © 2017 Michael Sanders. All rights reserved, DACS.

In this discursive text, artist Michael Sanders and Professor of Fashion Thinking, Jonathan Faiers outline the nuclear history of Holy Loch and Gare Loch in Scotland and consider the role of art and textiles in critically engaging with these sites.

As the 'Blundering Nuclear Tourist', Sanders, dressed in a specially created suit made from *Polaris Military* tartan (see Figure 7.1), momentarily occupies nuclear

DOI: 10.4324/9781003622185-8

sensitive sites to make visible the invisible nuclear resonance of these locations. The *Polaris Military* tartan, along with *US Navy Edzell* and the restricted *HMS Neptune* tartans, joins Sanders' own *SALT3 Submarine* tartan as textile manifestations of the infiltration of nuclear culture into everyday life. Tartan, which includes in its own complex narratives, histories of oppression, subjugation and territorialisation, becomes, in Sanders' practice, both camouflage and exposure. The deceptively attractive tartan pattern (or 'sett') of *SALT3*, for example, is determined according to the different quantities of radioactive elements present in the waters of the submarine base at Gare Loch in Scotland.

Who has the 'right' to wear a specific tartan is, of course, a significant factor in its own development and history. From the English attempt to banish it from Scottish Highland culture following the defeat of the Jacobite cause at Culloden, on through the nineteenth-century commodification of clan tartans and the branding of Scotland as part of a unified Britain, tartan's history has been both liberating and oppressive, celebrated and derided. As visible expressions of established systems of exclusion and authority, tartan makes evident that which often goes undetected, and just as the *Balmoral* tartan was designed to be worn only by members of the British Royal family in the confines of their estate, the previously mentioned contemporary *HMS Neptune* tartan is restricted only to those authorised by the Commanding Officer of HMS Neptune, the on-shore establishment of HMNB Clyde, or Faslane, as it is more commonly known.

This chapter situates these 'nuclear' tartans and Sanders' performances within nuclear cultural discourse, utilising them as co-ordinates for today's nuclear landscape, one that is both culturally and physically contaminated. In this space 'contaminated' tartans and those who wear them become radiant objects, illuminating the relentless processes of occupation that today's nuclear presence continues.

Contaminated Tartan

Michael: In March 1960, at Camp David, President Dwight D. Eisenhower and Prime Minister Harold Macmillan agreed to base US nuclear submarines on the West Coast of Scotland in exchange for the sale of Skybolt nuclear missiles. The decision bypassed the involvement of the Scottish Office; indeed, Macmillan's Scottish ancestry was arguably the only Scottish element in the whole agreement. On 3 March 1961, the Polaris submarine tender USS *Proteus* sailed into Holy Loch, with the first submarine arriving shortly after, starting an American naval presence which would persist until March 1992. The base would serve the submarines of Submarine Squadron 14 (Subron 14). It was eventually joined by the dry dock USS *Los Alamos*, and a variety of tender ships would come and go over the years. The submarine tender and dry dock crew numbered about 1,665, not including the crews of the visiting submarines. Many crew members would be joined by their families, and their children attended local Scottish schools.

Dunoon, the local town adjacent to Holy Loch, was a small seaside resort with a population of about 14,000 people. The siting of the base meant a risk of nuclear accidents, becoming a prime target for Soviet attack and the focus for anti-nuclear protests. It also brought a hope of prosperity and the noise and disruption of a naval base to the town and, over the years, some of the trappings of naval life, including bars, the odd brothel, illicit drugs, one small riot and American cars.

'The US Navy made great efforts formally to integrate. Many social events were organised, and clubs and societies encouraged' and '[h]undreds of marriages took place over the years and the personal bonds brought a network across the US and Scotland which continues today' (Messersmith 2003, pp. 40, 119). This presence and its effects I have come to realise are certainly more nuanced than my first impression of an occupying nuclear force, and indeed, I have heard it said – 'they protested when they came, and they protested when they left'. However complex the ramifications of nuclear occupation have proved to be, the use of tartan could be interpreted as a sign that the Americans felt at home in Scotland. In the 1980s, RAF Edzell, home to the Naval Security Group Oceanographic Monitoring Station, which monitored Soviet Navy and submarine activities from Edzell, marked its territory on the east coast of Scotland again registering its own tartan *US Navy Edzell*.

Alongside the possible benefits of American culture, the presence of SUBRON 14 also brought a worrying side effect in the form of radioactive contamination. In the words of Captain Art Bivens:

> When a nuclear submarine starts up its reactor the coolant water heats up and expands. In the early days of operating these nuclear plants we simply discharged this water into the surrounding water and figured it was harmless because of the dilution effect. However, the British authorities called us on that assertion when they took bottom core samples underneath the tender in the Holy Loch and the samples turned out to contain minute corrosion particles of radioactive Cobalt-60.
>
> (Messersmith 2003, p. 103)

Furthermore, as investigative journalist Duncan Campbell (1986) writes:

> The submarines were regularly discharging excess coolant water, in which radioactive isotopes - especially Cobalt-60 - had accumulated, into the Loch. The discovery of unexpectedly high radiation levels was admitted by the Ministry of Defence in August 1965. Their public statements were intentionally couched in reassuring terms, but nevertheless confirmed that a substantial radiation leak had gone undetected until at least four months before.
>
> (p. 217)

This was to be the start of a pattern of downplaying risk and the adoption of a patronising attitude to local concerns that exists to this day concerning the presence of US and UK nuclear submarines in Scotland. The lack of local consultation and a tendency for vital decisions to be made in Westminster has fuelled the desire within many Scots to have more control over their own affairs.[1] Most seriously, US submarine bases on foreign shores appear to have carried out practices which would not have been acceptable at domestic ports. Campbell (1986) again notes:

> Yet such incidents continued and have been documented at Holy Loch's sister base in the Pacific, Guam. In 1975, the depot ship there dumped highly radioactive reactor cooling water into the Pacific, causing radiation levels at nearby public beaches to rise to over 50 times the allowable dose. (p. 217)

The lack of consultation with local people points to an exploitative relationship between the military and civilian and indigenous peoples, characteristic of nuclear colonialism (Kuletz 1998).

Jonathan: There is no denying that tartan is a radiant object. Brightness, vibrancy and intensity spring immediately to mind when searching for terms to describe tartan's conspicuous grids. This common understanding of radiance as synonymous with brilliance and visibility belies the concealed, we might even say invisible, histories woven into tartan's complex fabric. This duality, at once highly visible, yet imperceptibly pervasive, forges what initially might seem a dichotomous alliance – that of tartan and nuclearity. While visual manifestations of nuclear culture are familiar in film, photography and other media as well as its physical occupation of space in the form of architecture and the demarcation of landscape, the actual effects of nuclear material remain intangible, operating in a space undetectable by the senses. This play between the tangible and intangible, the covert and hidden, is both exploited and interrogated in the work of Sanders, whose practice utilises the visible manifestations of nuclear culture, specifically the USA's nuclear presence at Holy Loch submarine base and what might be understood as its associated 'nuclear' tartans, to expose the unseen yet pervasive effects of nuclearity. For Sanders, tartan is uniquely positioned to act as both conduit and medium through which to address concerns central to his practice, including the concepts of 'cultural insensitivity', and an understanding of 'camouflage'. The seductive surfaces of tartan act as a form of camouflage both literally when worn in his performances but also metaphorically, imitating the camouflaging of potential destruction and long-term contamination inherent within the nuclear occupation of the sites Sanders interacts with.

Michael: I first chanced upon *Polaris Military* in 2007. The tartan was commissioned in 1964 by the officers and men of the American submarine base at Holy Loch in Scotland and was the first ship and nuclear weapons system in history to have its own tartan. The Scottish Register of Tartans explains that: 'The Lord Lyon of the time apparently exploded when the tartan was submitted to him and is quoted as saying: "We don't record tartans for submarines. It is nonsense. I have never heard of a ship's tartan in the whole history of Scotland!"' (*Polaris Military*. The Scottish Register of Tartans). Despite this, the tartan was registered, and *Polaris Military* tartan is still worn today by the Pipes and Drums of the US Naval Academy. Accepting that tartan's versatility means everyone has a right to claim it, is there not something obscene about the creation of a tartan as a symbol for a weapon of mass destruction?

In 1992, my friend, Mark Lumsden Cook, showed me some beautiful tartan baseball caps he had commissioned as gifts for his brother's wedding. Mark wanted something to welcome his American in-laws into the family. The bride's father was a Four-Star General, a decorated veteran of three wars: WW2, Korea and Vietnam. Mark said: 'the General's artillery unit had been tasked with the job to shoot the swastika off the top of the Nuremberg Arena! He was quite a man!' The baseball caps were made in *Lumsden Ancient* tartan and were perhaps an attempt to use tartan's powers to conjure instant tradition[2] with which to impress his American in-laws.

I stored this in my memory and resurrected the idea when I came to use the *Polaris Military* tartan for a tartan suit. I was preparing work for a joint exhibition[3] that included work about the accidental damage to archaeological sites at Babylon after the second Gulf War, when an allied military camp was built there, and sandbags were filled by digging into the archaeological site. The exhibition raised questions as to how we might protect high-level nuclear waste sites far into the future, if we failed to protect a known heritage site now. The *Polaris Tartan* suit was an acknowledgement that I was behaving like a naïve tourist but also utilised a degree of ironic, self-effacing humour to act as a foil to the seriousness of the subject matter. Additionally, I wanted the suit to operate as both a critique of cultural insensitivity and as the livery of a nuclear pilgrimage, drawing on the way in which tartan, as a cultural icon of Scotland, was appropriated as part of this nuclear occupation, culminating in the *Polaris Tartan*. Exploiting the familiarity of tartan is an effective way to infiltrate and communicate complex ideas into the everyday.

Jonathan: The sett of a tartan pattern takes the form of unique and individually designed grid patterns composed of different widths and colours. A grid structure can both contain and restrain but can also be considered endless or infinite. Physically, a grid composed of intersecting lines is only limited by the margins or edges imposed upon it. Without these limits its lines transcend borders, connecting with other lines from other grids to form new, ever-expanding networks. Similarly, nuclear material does not recognise borders; its dispersal and contamination are extensive, rapid and largely unseen. This integral duality of the tartan grid is fundamental to an understanding of the complex socio-cultural history of tartan, as a cloth and pattern that has, in turn, been liberating, defiant, oppressive and colonising. It has been, and is, worn both as a symbol of resistance and as a livery of subjugation. Tartan has a uniquely complex and often uncomfortable history, in particular tartan's many associations with the processes of colonial expansion and subjugation, a contributing factor to its undeniable multi-valency makes it a particularly relevant object with which to consider the American nuclear presence in Scotland. Tartan's adoption by the American forces stationed in Scotland is only a more recent example of the cultural insensitivity that structures tartan's ongoing legacies of colonisation and assimilation.

The *Polaris Military* tartan functions as a fragment through which the hyperobject of radiation can be perceived. This tartan is a form of textile agent that acts simultaneously as transformative on both wearers and viewers, but whose covert historical power, camouflaged beneath its dazzling, radiant surfaces, also imbues it with a powerful, largely unobserved force which is hidden in plain sight, echoing the effects of nuclear contamination. Sanders' use of tartan mirrors what we might understand as the American deployment of tartan as a form of a textile Trojan Horse, using Scottish tradition in the form of tartan to gain acceptance while simultaneously, under its familiar textile guise, infiltrating the local area with its nuclear load. Sanders deploys tartan's problematic history, but also its seductiveness, its 'softness', its very 'tartan-ness' is a form of camouflage, which *he* hides behind (or Trojan Horse-like, within) in order to introduce the inherent dangers and pollution of nuclearity to audiences who are initially seduced by tartan's superficial beauty. This usage goes beyond the mere harnessing of tartan's global recognition and appeal. So much of Sanders' work, especially the original tartans he is producing, references the fact that many tartan patterns, their distribution and proportions of different colours reflect specific geographical and or cultural references. Many tartans and their colour palettes profess to be inspired by the locales which they commemorate, the grey background of Balmoral tartan being a case in point, produced by threads of black and white yarn to imitate the rough-hewn granite so familiar on Royal Deeside.

Michael: In 1962, the US cancelled the Skybolt missile programme and following a series of negotiations the UK opted for a seaborn nuclear deterrent. In 1963, the US agreed to supply Polaris missiles and some submarine propulsion technology. The upgraded Faslane submarine base on the Gare Loch opened in 1968 along with an arms depot on Loch Long at Coulport and refit depot at Rosyth outside Edinburgh. In 1968, the first of four Resolution class submarines carrying Polaris missiles went into service. These were replaced in 1993 by Vanguard class submarines which carry Trident nuclear missiles. Like the nearby American submarine base at Holy Loch, the Royal Navy establishments continued the practice of discharging radioactive effluent into the coastal waters. A difference being that the waste was discharged through pipelines into the open waters (RIFE 26 2020, p. 232). (Also Devonport naval base's nuclear role – CND accessed online 6/10/2025, p. 4)

While I have taken the *Polaris Military* tartan to sites around the world, the first time I returned it to its place of creation in Holy Loch in Scotland was in 2017. To accompany the tartan suit, I also wore some nuclear tourist casual wear. Featured amongst it is a gold work blazer badge which declares, *We Pump Unseen*, a reworking of the Royal Navy's submariners' motto, 'We Come Unseen'. The Royal Navy at HMNB Clyde, commonly known as Faslane, routinely pump radioactive effluent including Cobalt-60 and tritium into the Gare Loch. The isotope tritium is also vented to the air from the nuclear weapons stored at Royal Naval Armaments Depot (RNAD), Coulport. This can make its way into the local environment in 'tritiated' rainfall; a contamination of the environment that affects all those living in it, including those responsible for the contamination.

As a development of the project *We Pump Unseen*, I considered ways to further highlight the issue of radioactivity from the submarine bases. As outlined, for many years reactor coolant water containing radioactive isotopes has been disposed of into the sea using dilution to wash it away. There is no such thing as 'away' and the contamination finds pathways up the food chain. Using salt making (which has its own long tradition as an effective means of protest), I planned to make a small quantity of potentially inedible table salt using this contaminated seawater to reconcentrate the pollution. In 2021, I registered the trademark for a table salt *SALT3* and started considering designs for a new tartan to accompany it. The table salt and tartan fabric is intended to make tangible the dangers of radiation which cannot be seen, smelled or tasted. The half-lives of the radioactive isotopes mean this table salt will have a 'best after' rather than a 'best before' date.

Around this time a bizarre echo of the past resurfaced – I noticed press reports that Faslane/HMNB Clyde had commissioned its own tartan, with the creator, Brian Wilton, joking that it could be 'glow in the dark' (Christison 2015). An article in the *Daily Express* stated:

> Captain James Hayes said: "We are very proud to be the only naval base in Scotland and so wanted to highlight our Scottish identity by creating the *Neptune* tartan. The chosen design incorporates HMS Neptune's location in the Scottish landscape, the unit's history and also the globally recognisable colours of the Royal Navy".
>
> (Christison 2015)

The design for *SALT3 Submarine* tartan is intended as an answer or antidote to the *HMS Neptune* tartan, possibly for use in ceremonial acts of salt making and to challenge the restrictions placed on the military tartan, whereby only those authorised by the Commanding Officer of HMS Neptune may wear/use/weave this tartan. Aware that I risk re-enacting the original insensitivity of the *Polaris Military* tartan, nevertheless, the process of registering a tartan with the Scottish Register of Tartans is alluring. In order to affirm its unique thread count, the process encourages the provision of a narrative or rationale for new tartans, making tartans into fragments of conceptual art. For example, on the Register the *Polaris Military* tartan is described in the following way: 'The navy blue represents the naval uniform, the dark green the depths of the oceans and the Royal Blue and gold overchecks represent the 'Blue' and 'Gold' crews who alternate' (*Polaris Military*, The Scottish Register of Tartans).Accessed online 4/10/2025.

The *SALT3 Submarine* tartan signifies pollution around submarine bases in UK coastal waters, its design taking inspiration from submarines that must remain unseen but make their presence felt to present an effective deterrent. The dark ground is intended to represent naval bases at night and the acoustic tiles on Vanguard nuclear submarines; rust red references the decommissioned submarines stored at Devonport and Rosyth and neutron activation of iron in their nuclear reactors; blue, green and grey are intended to represent the pollutants Cobalt-60, tritium and carbon-14 created in reactions between uranium reactor fuel, its cladding and water; white denotes salt made from local seawater; red is intended to represent the patrols escorting submarines in and out of port and grey is a nod to the pinstripe suits worn in Westminster where nuclear decisions are made. The sett reflects the isotopes uranium-235, hydrogen-3 and carbon-14.

SALT3 Submarine tartan was originally conceived to incorporate luminescent or reflective stripes indicative of the radioactive isotopes woven into its story and to give the Ministry of Defence police a sporting chance if it were to be worn during clandestine acts of salt-making at night. Samples have been woven by independent weaver Jonny MacKinnon with reflective and UV-sensitive thread, allowing the tartan to be viewed in different ways in different lights.

Jonathan: Much of Michael's previous work utilising the existing *Polaris Military*, *US Navy Edzell* and *HMS Neptune* tartans concerns the transgression, testing and redrafting of borders. Both the metaphorical borders represented by the tartans' grids and the networks connecting transatlantic nuclearity and Scottish local communities, but also the borders of the nuclear sites themselves, borders encompassing the geographical, cultural, economic and social. His more recent work, however, including his own *SALT3* tartan, is more concerned with exploring the concepts of dilution and reconcentration made evident in the filtration processes that have engendered the conception of this new tartan. As befits the tartan's design and the performative actions attendant on its creation, *SALT3* focuses more intensely on the effects of nuclear contamination and specifically its dispersal in water. Tartan, in this latest body of work, functions as a dilutant, the cloth's visual appeal, popularity and resonance mitigating the deadly effects of the nuclear contamination of UK waters, a contamination that is attested to in the aforementioned thread count that makes up the sett of *SALT3*.

Michael: Future plans for *SALT3* include Highland dancers and pipers dressed in the tartan performing ceremonial acts of salt making while taking radioactive contaminated seawater from alongside three submarine bases in England and Scotland.

The seawater collected will feed a salt-making machine and textile monument made of tartan fabric. As the seawater evaporates, the monument will become crusted in salt. The monument is intended to be portable, a mocking memorial to the nuclear age and its unresolved issues of legacy waste. It will playfully provoke questions about the treatment of nuclear waste by asking us at what point does the contaminated seawater become nuclear waste?

Jonathan: Nuclear and textile material culture might on first consideration seem an unlikely alliance; however, Michael Sanders' practice exposes a shared similarity, in this instance, between the nuclear presence in Scottish waters, and tartan. Both have complex, divisive and pervasive histories, histories that are woven from concepts of colonialism, cultural migration and ambivalence. Both engender passionate debate and have been welcomed and repulsed in equal measure. But beyond their explicit manifestation of geographical occupation and local identity, Sanders' interrogation of the imbrications of nuclear and textile cultures helps us understand the devastating power and pervasiveness that lies unseen beneath the dazzling surfaces of these radiant objects.

Notes

1 A detailed consideration of this can be found in: Dobson, Alan P. and Whitham, Charlie, 'Local Consequences of a Global Cold War. Chapter 7 'Project Lamachus: The Cold War Comes to Scotland - The Holy Loch. U.S. Nuclear Submarine Base and Its Impact on Scotland, 1959-1974', pp. 169–192.

2 Instant tradition was a term used by Professor Hugh Cheape at the Exploding the Grid conference. V&A Dundee 2024 and I believe references Hugh Trevor-Roper's work, *The Invention of Tradition: The Highland Tradition of Scotland* (1983) (see References).

3 *Ruin: An Investigations of Future Archaeology* (2008). A collaboration with Walter Cotten. Kruglak Gallery. Oceanside, California (see References).

Bibliography

Campaign for Nuclear Disarmament. n.d. *Devonport Naval Base's Nuclear Role*. Available at: https://cnduk.org/resources/devonport-naval-bases-nuclear-role/ [Accessed 6/25].

Campbell, Duncan. 1986. *The Unsinkable Aircraft Carrier – American Military Power in Britain*. Second Edition. Great Britain. Paladin Books.

Centre for Environment, Fisheries and Aquaculture Science on behalf of, Environment Agency, Food Standards Agency, Food Standards Scotland, Natural Resources Wales, Northern Ireland Environment Agency and the Scottish Environment protection

Agency 2021. *Radioactivity in Food and the Environment, 2020-RIFE 26*. Great Britain

Christison, Greg. 2015. Faslane issues fashion statement with new military tartan trend. *Express*. Available at: https://www.express.co.uk/news/uk/562589/Faslane-issues-new-fashion-statement-Clyde-servicemen-and-women-military-tartan [Accessed 6/25].

Dobson, Alan P. and Whitham, Charlie. 2007. 'Project Lamachus: The Cold War Comes to Scotland—The Holy Loch U.S. Nuclear Submarine Base and Its Impact on Scotland, 1959–1974'. In Jeffrey A. Engel (ed.), *Local Consequences of the Global Cold War*. Woodrow Wilson Center Press/Stanford University Press, pp. 169–192

Kuletz, Valerie L. 1998. *The Tainted Desert – Environmental and Social Ruin in the American West*. New York and London: Routledge

Messersmith, Andrene. 2003. *The American Years – Dunoon & the US Navy*. First published in 2003 by Argyll Publishing. Great Britain: Paperback edition published by Capercallie Books Ltd under the imprint Argyll Publishing in 2016.

Radioactivity in Food and the Environment (RIFE-26). 2020. Available at: https://www.sepa.org.uk/media/593789/rife26.pdf. Accessed 10/25.

Sanders, Michael and Cotten, Walter. *Ruin. – An Investigation of Future Archaeology*. 2008. [Kruglak Gallery]. MiraCosta College, Oceanside, CA, USA. 4 March – 3 April 2008. Details available at: http://www.eastcomplex.com/ruin_info.htm [Accessed 6/25].

Tartan: Exploding the Grid. 2024. [V&A Dundee]. 12 January 2024.

The Scottish Register of Tartans. n.d. Available at: https://www.tartanregister.gov.uk/index [Accessed 6/25].

Trevor-Roper, Hugh. 1983. 'Chapter 2. The Invention of Tradition: The Highland Tradition of Scotland'. In Eric Hobsbawm and Terence Ranger (eds), *The Invention of Tradition*. Cambridge: The Press Syndicate of the University of Cambridge, pp. 15–41.

8 Remembering Plutonium

Tim Andreae

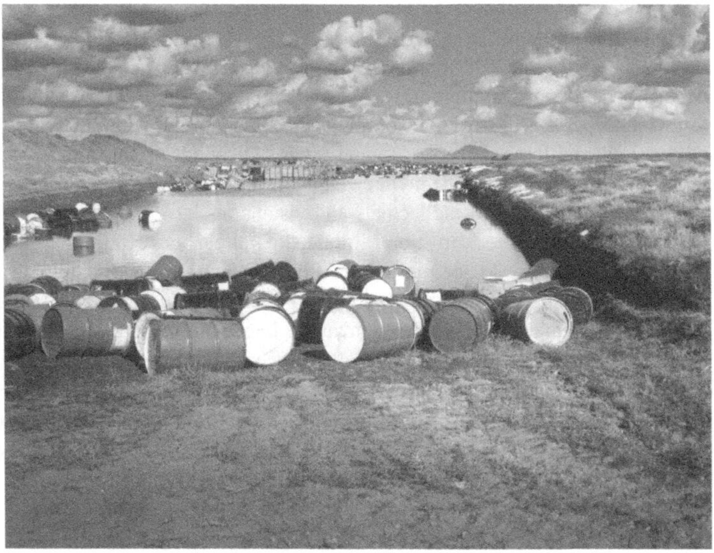

Figure 8.1 Floating barrels of radioactive waste at the Idaho National Laboratory (INL) ca. 1962.

Source: Image from INL archive (public domain).

In the late Summer of 2012, I went on my first tour of the radioactive "cleanup" sites at the Idaho National Laboratory (INL) (see Figure 8.1), one of a network of sites across the country run by the Department of Energy (DOE), formerly known as the Atomic Energy Commission, that branch of US government, devoted to the research and development of nuclear power and nuclear weapons. I had just joined the board of a small, grassroots, nuclear watchdog organization called the Snake River Alliance. I arrived at the site with no idea the place would become a touchstone for me, looming large in my psyche and in my creative

DOI: 10.4324/9781003622185-9

life for more than a decade and counting. From that first contact, I sensed a story too big to apprehend – not just the flood of technical information, or the mazes of billion-dollar facilities, or the military personnel walking around with machine guns, but something else, as if beneath all the shiny extroversion was an unthinkable weight. While I found the surrounding peaks of the Lost River Range and the sea of sagebrush beautiful in their starkness, the miles of industrial complex didn't make me want to go back. But the next year I went back. And then I kept going back. Certain realities I couldn't wrap my mind around – like the fact that we are creating substances like plutonium that will be toxic to all living systems for hundreds of thousands of years – as good as forever. Not only that, but from 1954 to 1970, massive quantities of plutonium-contaminated garbage were buried in great unlined pits on the site, which sits over the top of the Snake River Aquifer. Such realities were unintelligible to me.

As it happened, in 2012, I also began a career as a mental health counselor. And so, when I saw a symptom, I was primed to not just gloss it over, but to follow it deeper. Because of my work, I had become familiar with the writings of Carl Jung and James Hillman, trauma experts Donald Kalsched and Vivian Broughton, not to mention the mythic vision of Martín Prechtel, Joanna Macy and James Russell. Such intellectuals and visionaries have provided frameworks that I continue to learn from. I noticed parallels between the INL and the people who came to see me who had been running from psychological pain for longer than they could remember. Perhaps we've all been running from pain for longer than we can remember. I see it in myself, and I see it with my clients. They come into my office overflowing with what they've been trying to outpace. There are so many ways to avoid tending to what is inside of us, but as Carl Jung observed: what we don't make conscious of our inner world will confront us from without as fate (Jung 1970, pp. 70–71). Jung called this a psychological rule, and it seems to hold true for both the micro and the macro; for the individual as well as the people. From my early engagement with the site, I saw radioactive waste as the shadow side of our industrial military complex, and from my work as a therapist, I recognized the value of approaching realities that have been suppressed by the dominant faction. And so, by my third visit to the INL in 2014, I had begun bringing a cohort of artists along with me on tours of the radioactive waste management sites with the idea of attending to what I came to see as a manifestation of disassociated pain that had been buried in the ground more than 50 years earlier. I called the project *Holding What Can't Be Held*, and for ten years, I toured the site with one group and then another, and we would sit in contemplation with what most people want to forget.

When I first sat with the idea of plutonium, a substance that will be toxic for 240,000 years, and the reality that waste from milling the plutonium cores of nuclear bombs is now buried in massive quantities over the greatest treasure of our region – a body of water more voluminous than Lake Erie, underlying 10,000 square miles of Southern Idaho – it stopped me. I hadn't known I'd been running. But for a moment, I stopped. I had been imagining myself headed towards somewhere better. "Always something better ahead". I suppose this is the way thieves think – thieves or ghosts. Before we know it, we're no longer telling stories around a fire but imagine ourselves headed towards something better. Before I know it, I am thieving myself. Instead of experiencing the pleasure of a moment, I've tucked the pleasurable experience into my belt just to get on my way a little quicker. If I'm successful at maneuvering my way to the front, everything seems to be going swimmingly, but the hurrying itself feeds a systematic extraction from the present moment that is occurring simultaneously on such a large scale it's hard to imagine. Just so, our story of technological progress, with always the promise of something better, leaves us and the landscapes around us impoverished. Confronted with a substance whose toxicity extends nearly forever, paired with the necessity of attending to it for time frames I couldn't grasp, something in me clicked – no amount of running could ever outpace those 240,000 years.

Occupying 890 square miles of the Arco desert, the INL consists of nine distinct facility clusters, though I tend to divide the site into two main parts: the part that's trying to account as best it can for the egregious mismanagement of the past, and the part that's always trying to get somewhere better – on one side, the various waste management complexes and on the other, the research and development projects. On the research and development side, the reality of radioactive waste slips like water through a sieve, while at the various multi-billion-dollar "clean-up" sites, the slow, concerted effort continues without fanfare. I too would rather focus on a big idea than try to manage an impossible situation. Likewise, within the nuclear industry, the big money flows into the new big ideas or at least new versions of the same old ideas. The Snake River Alliance's former nuclear program director, Beatrice Brailsford, used to say that the R&D part of the lab "has more money than God". The pithy comment never failed to bring a chuckle. But then, like a riddle, it made me think. The nuclear endeavour hinges on the shattering of something too small to see, but everything about it is big. From its unimaginable destructive power to the industry's attempts to find an endless source of energy, to the extreme longevity of its waste products, it's as if the entire venture were trying to push beyond the human realm into the realm of the gods. Several years ago, our small group walked through the INL's National Reactor Innovation Center with a team of nuclear engineers, military personnel and one very well-paid "environmental" lawyer. Recently designated as the US hub for new reactor design, the INL is currently the US industry's best hope for a long-awaited nuclear renaissance. As the lead engineer of the Microreactor Applications Research Validation & Evaluation (MARVEL) micro-reactor spoke, he exuded a sense of buoyancy that reminded me of my own childhood flights of fancy. The man was palpably giddy with a vision of the MARVEL micro-reactor mass-produced in factories. When my friend asked him what might be done about the problem of radioactive waste, including the waste that keeps accumulating at reactors around the country with no solution in sight, the man stopped short. He finally suggested we put it into an "impenetrable canister" and "shoot it into oblivion".

I guess he could have said, "Let's just forget about the waste, okay?" Though maybe that *is* what he said in an engineering kind of way. And while such a statement could elicit derision or a sense of superiority in an environmental activist, the truth is, I'm in the same boat. Everything in me wants to overlook the reality of radioactive waste. Somehow it feels connected to the sense of desolation I'm always trying to outpace. The need of our age is so heartbreakingly big. In his essay "Oedipus Revisited", James Hillman notes that *Oedipus Rex* begins with a sick city – not just the people in it, but the city. He writes,

> For Oedipus – or for any one of us – to find one's father and the truth of oneself is not enough, because there is a murder in the order of the world, and the world and the soul are inseparable. That's why the longing in the play is so intense: the *world* must be restored, not only its men and women.
> (Hillman 2007, p. 163; Italics mine)

Like Oedipus, we too would like to believe the riddle of the Sphinx has an answer and if we could just say it, we'd be done with it and would then join that class of vanquishing super heroes who always win the day, like Harry Potter and Superman (not to mention all the comic book heroes born from nuclear radiation), but also Perseus, and on and on, going back to Gilgamesh and others whose monster-killing stories are a façade for the rape and pillage of a conquering civilization and more fundamentally for the subjugation of Nature Herself. If, upon realizing this, we were to strike such stories from the annals of myth, we would then be doing more of the same. Hillman identifies this tendency in the therapy room: "The solutions imagined by a patient for his illness belong to the image of the illness" (Hillman 2007, p. 165). When the sick city imagines the path to health, the path is just another expression of the sickness. We are a sick city. We are all in the process of trying to outspend God. It doesn't help to tell ourselves to slow down. When we approach the affliction in front of us, we have to remember it is equally embedded within us and has its origin in the way that we have come to think and perceive as a people.

As a civilization, we had surely been running before we dropped atomic bombs on Hiroshima and Nagasaki. But after 1945, the urgency of our flight increased a hundredfold. Instead of contending with the horrors of the bomb, we were smitten by a utopian dream. While the Jetsons zoomed effortlessly across TV screens throughout middle America, the vision of a nuclear-powered Eden was crystallizing in our collective mind's eye. The INL was founded expressly to fulfil that dream. It is there we set out to find what wonders could be accomplished with this new power. "Energy too cheap to meter" was the slogan they preached, and in 1951, the INL's EBR1 was the first reactor to create electricity. The prototype for the first nuclear-powered submarine was created at the INL, and less successfully, prototypes for a nuclear-powered aircraft were built and tested, though they never made it off the ground. But if there's a substance that could be thought of as the link between our world-destroying and our attempts at world-building, it would be plutonium. Plutonium is what we use to make the cores of our nuclear bombs, but the way we get plutonium is through nuclear fission.

To understand plutonium as tied to our collective shadow, it's worth noting that the word comes from Pluto, the Roman version of the Greek god, Hades, who presides over the land of the dead. It's also worth considering the zeitgeist in 1954, the year we started trying to make the waste from the Rocky Flats bomb plant disappear by burying it at the INL in pits as big as football fields. Nine years had passed since we dropped atomic bombs on two cities full of innocent people. The auto industry was booming like never before. The polio vaccine had just been discovered, and 1954 marked the first mass vaccination of children against polio. This was the advent of "fast food". In 1954, the first Burger King was built, and in 1955, the first McDonald's franchise opened in Des Plaines, Illinois, debuting its famed golden arches and its "Speedee" service system inspired by Henry Ford's automobile assembly line. Along with the mass eradication of disease, the nuclear age seemed to be ushering in new levels of convenience, comfort and freedom of mobility. A dominant faction of American society was feeling unbridled hope in a technological dream of progress that featured the primacy of the individual. As it happened, 1954 was also the first and only documented case of an individual being hit by an object that had fallen from outer space when a four-kilogram piece of meteorite fell through a roof, badly bruising Ann Hodges, who had been napping on her couch in Sylacauga, Alabama, establishing an indisputable link between human life on earth and so-called "oblivion".

While the meaning of such cosmic echo-location must be left to the great mystery, what is certain is that when a culture elevates the individual above all else, it not only gives rise to an increasingly superficial world view, but it loses the capacity to incorporate death into its cosmology in a meaningful way. Such a world view manages death the way we manage radioactive waste – by keeping it out of consciousness for as long as possible. No wonder then that in the 1950s and 1960s, when spring runoff made a giant pond out of the open pits of radioactive waste, standard protocol was to shoot the floating barrels to sink them.

As a parallel to his psychological rule mentioned above, Jung also observed that the point at which we stop telling the old stories is the point at which we step directly into them. When the "gods" go unacknowledged, they don't just go away, but instead become the driving force within a psyche or within a people, often manifesting as illness, neuroses and obsessions (Jung 1967, p. 49). In the same way that the myth of Oedipus turns on a single prophetic threat, the entire nuclear industry has been driven by existential threat from the start. Whether all-out nuclear holocaust or the climate crisis of today, the sense of imminent peril and the urgency behind it has provided the industry with both the justification for its existence and pushed its experimentation to heights of absurdity. They not only blasted massive nuclear-powered aeroplane engines across the desert on railroad tracks but they pushed reactors to criticality, blowing their radioactive cores across the desert (Stacy 2000, pp. 116–125). These days, the INL has test reactors that will simulate accident conditions, but the underlying psychology and its push against the odds continue. As I suggested above, at the INL, our engagement with physical matter seems to mirror what we have not wanted to face as a civilization, offering us a glimpse of our collective neuroses, which drawn to their extreme ends can be witnessed in distilled form like a Greek tragedy with its layers upon layers of irony; concise as an Emily Dickinson poem that once unpacked could become an alchemical treatise for the nuclear age. Such a treatise would involve not just Pluto, but a whole pantheon of unacknowledged forces of nature. But for the purposes of this essay, I will tease out the single thread of death and endings as they relate to the nuclear industry, especially the pairing and unpairing of heroes and death.

Above, I mentioned the glut of vanquishing heroes that our culture has elevated, along with that dream of individual comfort. But in his essay, "An Epic for the Borderlands", Harvard professor and scholar of Iranian and Armenian mythology, James Russell describes a different kind of hero – one with more ancient roots than the ones mentioned above. From Vīštāsp, in the Iranian epic of the Kayanians, who chooses death over eternal life in a magical bronze fortress, to the Indian 'Mahābhārata' in which Arjuna is counseled to fight even though the battle will mark the end of a world-age, Russell argues that epic heroes grapple with death, typically succumbing to it, and it is this "sense of ending" that makes the tale an epic (Russell 2004, pp. 951–952). He then goes on to point out a shift reflected cross-culturally. He describes the ancient pagan epic of the Narts, whose story is paralleled more recently by the Armenian epic of Sasun. These stories portray a long line of heroes arriving generationally at protagonists, who instead of honoring the reality of death, try to circumvent it and as a result, are cursed by their fathers to live inside of a mountain until the end of time, neither living nor dying. Russel writes, "Violating the nature of time, they become unnatural. Clinging more to life than man must, they receive ironically the penalty of an endless life not worth living" (Russell 2004, p. 976).

In the Ossetian epic, "The Narts", Artavazd, rebels against his father's death rites, an act that condemns him to be chained inside of Mount Ararat and to never see the light, only to break loose in the apocalyptic age. Russell writes,

> Down to the high Middle Ages, smiths would strike their anvils thrice on the first work day of the week or else during the celebration of Navasard, the New Year, to strengthen the chains of the imprisoned apocalyptic prince; thus a religious element of the pagan epic survived at least a millennium in folk memory.
>
> (Russell 2004, p. 953)

Centuries later, in the Armenian epic of Sasun, the hero Mher, who in his flight from death not only suffers a loss of vitality and is rendered infertile but is likewise cursed by his father: "May you enter and remain in Carven Rock (Raven's Rock…), may you not see the world, deathless Cain, may three crumbs be your daily food: I have invoked God" (Russell 2004, p. 957).

And so, the engineer's fantasy of the impenetrable canister finds precedent in these rock caves that become the forever home of heroes who have spurned death. The patterning that got us to where we are now is not new. Imprisoned in rock for their hubris, these apocalyptic princes are a distant echo of the material by-products of our nuclear striving, for they both suffer the same curse – to live on, down through the ages, burning with malcontent, imprisoned, whether in concrete or in a deep geologic repository until the end of time. And who's to say that this work of nuclear fission, shattering the deep connective force within matter, isn't the apocalyptic prince himself breaking out of his imprisonment within Mount Ararat – the very same mountain where the biblical Noah landed the Arc to begin a new world age after the great flood?

Centuries after the Nart Sagas, Sophocles' *Oedipus Rex* could be imagined as tracing that same rupture of heroic ethic, as it turns on itself and compounds into full-blown sickness. Here, Oedipus' father Laius repeats a motif that appears at the founding of numerous civilizations, including Rome: he tries to kill his infant son rather than risk being killed himself when that child comes of age. But how does such a child come of age with a father who had never fully grown up himself? Russell also describes the Greek play of *Alcestis*, which tracks the plight of Admetus, who, upon facing death, is engaged in his own form of running. According to Russell, "The man who chases about asking parents and relatives to die for him, that is, to assume his burden of the human condition, is resisting adulthood, part of which is the recognition and acceptance of one's own mortality" (Russell 2004, p. 975). Referring to Admetus and others, Russell suggests that such "naïve struggles with angels… exemplify the unripe aspect of man's cultural youth" (Russell 2004, p. 976).

Having left the heroic age well behind us, we come to our current age where the values of the collective have devolved into childish, "me-first" fantasies; where those who resist the "burden of the human condition" are not cursed, but exalted. From the medieval tradition of alchemy, there is a cautionary image of a green lion swallowing the sun (Edinger 1994, p. 65). While its interpretations could fill volumes, the image would seem to express an unnatural reversal, for that which is "green" or unripe is swallowing the sun that brings everything to ripeness. But what about the lion? In An Epic for the Borderlands, Russell traces the symbolism of the lion from early Mithraic fraternities in which the lion embodied death, to the lion as a royal symbol of a king's power of life and death over his subjects. He remarks, "One cannot emphasize too strongly this symbolic linkage of the lion to death, and to man's confrontation with it" (Russell 2004, p. 955). When on a cultural level, the fact of death goes unacknowledged, the capacity for ripening into maturity is itself "swallowed" and the living become ghosted.

Far from confronting death, the heroic impulse of today would have us devise ingenious ways to get around it. Nuclear engineers, after all, are not seeking renewable energy, but they are seeking energy that *never dies*. With its dreams of an endless wellspring within matter itself and of aeroplanes and submarines that could go and go and go without having to refuel, since the beginning, the INL has been trying to bypass the death part of the life cycle, bringing to fruition an ancient curse – one that is no longer just a story of an apocalyptic prince living down through the ages chained up in a cave, but a story that by going *untold* for centuries has concretized into physical substance – substances now unleashed from the heart of matter, and that, for 36 years and as if by fate, were sent in massive quantities to the INL for disposal.

How did the apocalyptic prince break out of Mount Ararat, unless for the fact that we ceased to tell his story; ceased to offer even the three promised crumbs? Here, in our un-storied world, I find myself reaching for a cultural solution in the same way that nuclear engineers reach for a technological one. Identifying my grasping and the urgency behind it, I can't help but touch the place in me I want to fly away from. And I think of something Martín Prechtel has said: "If the world's going to end, what good is it to hurry? If it's not going to end what's the big hurry?" (Prechtel 2021, p. 166). When I stop hurrying, I have to contend with grief. The cultural frameworks that once nourished a soul-level coexistence with the natural world are nowhere around me. The bread crumb trail is gone, and we can't find our way back by reenacting ancient death rites or contriving to create new ones. The upshot – a collective soul-loss – echoes throughout our civilization's infrastructure, from interstate highways to strip malls to tract homes. Doggedly equating convenience with progress, we try to fill the cultural void with technological solutions – nuclear representing a kind of apex. Meanwhile, the toxic biproducts of that endeavor suggest the impossibility of untying our souls from the soul of matter. We cannot fly away from this place. To sidestep my grief at the situation is to leave behind the very thing that has generative potential.

In Krista Tippett's interview with long-time civil rights activist Ruby Sales, Sales centres the dialogue around a single fundamental question. She asks, "Where does it hurt?" (Tippett, 2016). Despite what the dominant paradigm would have us believe, it is the vulnerable place of relationality and human limitation that generates new possibilities, *not* the drive towards an efficiency-based, world-building, world-destroying, god-realm. Nuclear activist Joanna Macy wrote, "By choosing to honor the pain of loss rather than discounting it, we break the spell that numbs us to the dismantling of our world" (Macy and Johnstone 2012, p. 79).

In the Mahābhārata, Krishna, occupying the seat of the chariot driver, tells Arjuna the reason to fight is not for a particular outcome, but "to fight" is how Arjuna must serve. Such action, born from clear-eyed reckoning and lead by a love of something greater than oneself, is its own refuge. And who knows, perhaps lead by a love for the world and the courage to suffer the pain that connectivity brings, we could move towards something new, though as Martín Prechtel has said, as long as we include "ourselves" in our visions of world renewal, it will lead to more of the same. Instead, he says, we must feed "a time of hope *beyond* our own" (Prechtel 2025). To work towards a reality knowing we and our loved ones may never get to see it, is to work ourselves more deeply into the present. Then, like the epic heroes of old, we could become the rich humus from which something new and worthwhile emerges.

References

Edinger, Edward F. 1994. The Mystery of the Coniunctio, Alchemical Image of Individuation. Toronto: Inner City Books.

Hillman, James. 2007. *Mythic Figures*. New York: Spring Publications.

Jung, Carl. 1967. *The Collected Works of C.G. Jung, Vol. 13, Alchemical Studies*. Princeton, NJ: Princeton University Press.

Jung, Carl. 1970. *Aion: Researches into the Phenomenology of the Self*. 2nd ed. Princeton, NJ: Princeton University Press.

Macy, Joanna and Johnstone, Chris. 2012. *Active Hope, How to Face the Mess We're in without Going Crazy*. Novato, CA: New World Library.

Prechtel, Martín. 2021. *Rescuing the Light*. Berkeley, CA: North Atlantic Books.

Prechtel, Martín. 2025. Lecture XX – Courage and Blessings. Podcast: *Please Come Sit By My Fire*. Information available at: https://floweringmountain.com/pleasecomesitbymyfire/ [Accessed 7/25].

Russell, James R. 2004. *Armenian and Iranian Studies, Harvard Armenian Texts and Studies, 9*. Cambridge: Harvard.

Stacy, Susan M. 2000. *Proving the Principle: A History of The Idaho National Engineering and Environmental Laboratory 1949-1999*. Idaho Falls, ID: Idaho Operations Office of the Department of Energy.

Tippett, Krista. 2016. Ruby Sales. *Where Does It Hurt?* Podcast: On Being. Available at: https://onbeing.org/programs/ruby-sales-where-does-it-hurt/ [Accessed 7/25].

9 More-Than-Human Life and Art in Low-Level Radiation in Fukushima

Trevor Paglen *Trinity Cube* and *Don't Follow the Wind*

Jason Waite

This moment of planetary crisis, consisting of immense carbon outputs, mass deforestation, oceanic trash islands visible from space, and microplastics turning the inside of human and more-than-human bodies into landfill, is setting us on a course for a mass multi-species extinction event. In order to change this drastic outcome, we need more than just a shift in consumption; a large-scale ontological shift in how humans co-exist with our fellow inhabitants on this planet is necessitated. How can art provide a testing ground for this radical cultural shift that is needed?

One place to look for examples of transformative models is in sites of catastrophe that have already undergone massive environmental destruction and change. In these more extreme planetary areas, we can perhaps portend future scenarios for how the earth might look more broadly, and at the same time experiment with reparative cultural approaches that might stave off planetary collapse. One such model site is the Fukushima exclusion zone in Japan, formed in the wake of the triple meltdown of nuclear reactors in Tokyo Electric's Fukushima Daiichi Nuclear Power Station that began on March 11, 2011. This ongoing nuclear disaster has displaced tens of thousands of people and released a massive number of radioactive particles into the atmosphere, oceans, and across the earth's surface. The highest concentration of nuclear fallout from the disaster is in an area adjacent to the nuclear power plant, traversing seven towns and over 300 square kilometres—equivalent to an area larger than the city of Edinburgh. This dense area of contamination forms the Fukushima exclusion zone, deemed by the government to be too toxic for human inhabitation, leading to the forceful displacement of its human population and the area being closed to the public.

In response to this ongoing catastrophe, a collective of artists, curators, and cultural workers set about to create a long-term artistic project inside the Fukushima exclusion zone. This project, entitled *Don't Follow the Wind*, worked with displaced residents to host a series of new commissions inside the zone by artists and art groups in homes where residents could no longer

DOI: 10.4324/9781003622185-10

live and places of work no longer accessible (Hirsch and Waite, 2021). The exhibition opened on the fourth anniversary of the disaster on March 11, 2015, however, as the zone is closed to the public, the exhibition is waiting to be seen at some point in time in the future when the residents can return to live in the area. The exhibition is both an act of solidarity and a way to work on the critical imagination of all of us living outside of the area to show how art can act as a beacon for durational disasters.

One of the artworks in the exhibition is by artist Trevor Paglen, whose practice has long investigated the US military industrial complex. Paglen has worked to make visible much of its physical infrastructure, which remains remote and out of the public spotlight. His project for *Don't Follow the Wind* looks at the origins of the nuclear project in the United States, as well as acting as a material witness to the ongoing disaster in Fukushima. Entitled *Trinity Cube*, the artwork is made up of two materials. The first is trinitite, the human-made new element created in the midst of the first atomic weapons test that took place in New Mexico on July 16, 1945, codenamed 'the Trinity test'. The resulting nuclear fallout from the nuclear test seriously affected the health of indigenous people and rural residents in the surrounding area, as well as many of the soldiers present. The force of the nuclear blast fused radioactive plutonium with bomb debris and desert sand to form trinitite, a new element that can still be found around the test site today.

The second material in the artwork is shattered glass that was collected around the Fukushima exclusion zone, which also had radioactive elements on it from the nuclear fallout from the Fukushima disaster over half a century after the initial atomic bomb test. As theorist Sabu Kohso (2020, p. 89) notes, so-called civilian uses of nuclear power and nuclear weapons are "Janus-faced"—two deeply connected facets of the nuclear project. The artist fused together the two materials to form a 30-centimetre square cube that Paglen titled the *Trinity Cube* after the nuclear bomb test that helped to catalyse the nuclear project. The object was then placed inside the zone—reinterred in the contaminated environment wrought by the nuclear project itself. While as an artwork, *Trinity Cube* has an immense symbolic value, existing in Fukushima, the artwork also has a more subtle aim. As the zone had forcibly evicted its human residents, *Trinity Cube* can exist in a place where no humans can live. This means that the artwork can act as a material witness, testifying to the ongoing displacement in the area as well as by being situated there, potentially gathering material evidence in the form of radionuclides on its surface to document the ongoing disaster. Thus, the artwork acts in the same capacity as an aeroplane black box during a time of crisis—providing an uncensored recording of the ongoing social-environmental catastrophe in the area. This durational capacity of the artwork means that *Trinity Cube* can remain for years or decades in the radioactive zone. This integrity is important as *Trinity Cube* is not located in the controlled environment of a museum, with its multiple levels of security and conservation, but rather it is situated in a structure with no electricity, temperature regulation, or conservators. Rather, the artwork is exposed to what theorist Rob Nixon (2011) calls the 'slow violence' of radiation. The imperceptible and, at times, intergenerational violence undertaken at the molecular or genetic level is often not even perceived as violence. *Trinity Cube* both collects the evidence of this slow violence on its surface in the form of radionuclides and can withstand their energetic onslaught.

Years after *Trinity Cube* was installed in the exhibition in the zone, the artwork had an unexpected visitor that changed the perception of how this slow violence functioned and drew attention to how art can be a bridge, connecting different understandings of durational disaster as well as fostering relationships between humans and their environment.

Interspecies Infrastructure

Two years after *Don't Follow the Wind* had opened, members of the eponymous collective that organised the exhibition were still tending to the show multiple times a year, undertaking maintenance work on the art as well as the structures. During one of these visits by the collective, it was discovered that one of the sites of the project, an artist studio, which held an artwork by Taryn Simon, was broken into. This shocked the collective as no humans were living in the area; however, they soon discovered that it was a wild boar that had continually rammed the metal door to the studio until the door was disfigured and opened. The boar then rummaged around the artist studio until it found oil paints that had been left in the studio, biting into them and then spreading the black paint around the studio floor in a gestural manner reminiscent of abstract expressionism. This act by the boar of contributing their own "work" to the exhibition shocked the collective and forced them to reconsider the public around the exhibition. While the collective had been focusing on the plight of the displaced residents and the general public, which was prevented from seeing the show because of the limits of the exclusion zone, this contribution to the exhibition by the wild boar made the collective realise that there was an entire more-than-human public which the group had not been considering as part of the community around the show.

In order to find out who this more-than-human community was, and how that community was using the farm, the collective *Don't Follow the Wind* began to collaborate with scientists, researchers, local foresters, and hunters. As no human could live within the zone, remote sensing field cameras were installed on the site, as well as newly developed field microphones. The images and sounds that began to emerge from the site revealed that there was a more-than-human community that was thriving there. Wild boar, whose population had grown at a rapid rate since humans were forcibly evacuated, were routing around the site finding sustenance in the roots and insects not the sites (Saito et al., 2021). Moreover, animals that had not previously been seen in the area, such as macaques and Japanese serow, appeared and seemingly began to move into the area. What can be understood from this diversifying and thriving more-than-human population is that the previous human stress on these populations was more impactful than the radiation now present in the area. This is not meant to idealise the situation or frame the exclusion zone as a post-disaster arcadia; a characterisation has been common in the scientific literature around Chernobyl (Dombrovski, Zhurauliou, and Ashton-Butt 2022).

The exclusion zone in Chernobyl and the exclusion zone in Fukushima share very different futures. While the zone in Chernobyl was evacuated and has been left unoccupied for the foreseeable future, the zone in Fukushima has been undergoing large-scale remediation for a human population to eventually return to the area. This means that in the near future, humans will be returning to the area that, in their absence, has given way to a large and thriving more-than-human population.

This unique dynamic of humans moving back into a post-nuclear disaster zone with a large-scale more-than-human population provides an important opening for humans to renegotiate how they co-exist with their more-than-human neighbours and the environment more broadly. As such, Fukushima is an exemplary testing ground to consider how humans can transform their culture and to move towards a more interspecies way of understanding their surroundings and co-existing with them. Towards this end, the artworks in *Don't Follow the Wind*, including Trevor Paglen's *Trinity Cube*, have begun to undertake a new significance.

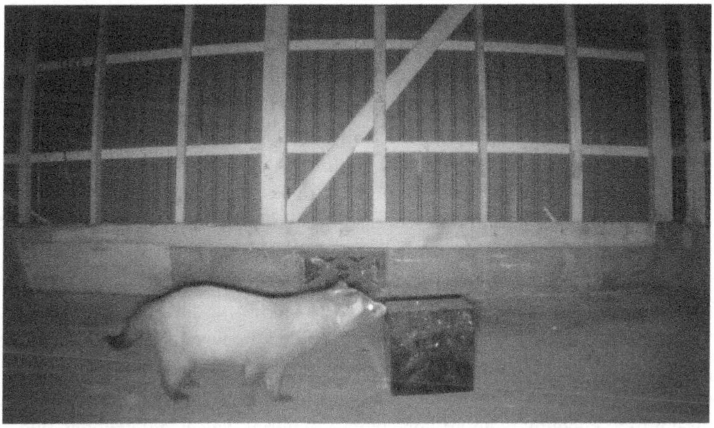

Figure 9.1 *Don't Follow the Wind*, image of the first visitor to Trevor Paglen's *Trinity Cube*, 2023.

Source: Courtesy of *Don't Follow the Wind*.

The field cameras, installed by the collective in collaboration with the researchers and scientists, and that have looked out at the surrounding environment, have also been turned inward to look at the artworks to see how this growing more-than-human community is perceiving the artworks in their midst. To this end, at twilight on October 8, 2023, a field camera facing *Trinity Cube* recorded what might have been the first more-than-human visitor to the artwork, (see Figure 9.1). *Trinity Cube* was situated on a concrete floor in a vacant warehouse that, before the disaster, was utilised to store charcoal. A field camera was later installed in the room at the same height as the artwork. In the resulting twenty-second video that was captured by the camera, a masked palm civet appears on the right side of the room, cautiously approaching the artwork. Leading with its nose, the civet sniffs the top right corner of *Trinity Cube* and then passes it by. While we do not know what this first visitor thought of the artwork, there is still much we can learn from the encounter. Firstly, what is clarified in this encounter is the multiple ways in which beings navigate the world, while as humans we often navigate our environment through sight, the video highlights how the masked palm civet approached the *Trinity Cube* with its nose. This "smell first" hierarchy of senses underscores the diverse ways in which beings can apprehend the world, highlighting how different senses can be used to apprehend an artwork. As researchers Marianna Szczygielska et al. (2022, p. 13) ask in their collaborative text *Feral Ethics*, what if "we started thinking with our 'snouts'". This sensorial provocation highlights the direct relation between thinking scents and diverse forms of cognition or even contemplation. Thinking with our snouts "smells" how communication, whether that be in the form of an artwork or being-to-being communication, need not only rely on spoken or written language but also on olfactory discourses. If the present planetary crisis underscores the need for an ontological shift in how we apprehend the world, then we can see how we need to consider the wider sensory world, including the epistemology of smells, as a means of both communication and knowledge production.

The question of who the guest is and who the host is, with relation to Trinity Cube and the masked palm civet, raises further questions relating to indigenousness and durational connections with the land. Both the masked palm civet and *Trinity Cube* ended up in Fukushima due to World War Two. The civet, which is indigenous to South and Southeast Asia, was brought to Japan by humans in large numbers to be farmed for fur during World War Two (Masuda et al. 2010). Those civets that survived the war and escaped from the farms have thrived in the environment of Japan and spread throughout the archipelago. Hence in this meeting the civet, brought to Japan as a consequence of conflict, encounters an artwork whose material was also developed due to the war and the civilian use of nuclear power. That this material was also a means for the military industrial complex to continue researching and designing nuclear weapons directly ties this four-legged mammal and the artwork together.

As part of developing a new mutual relationship between human beings and more-than-human beings moving forward, we need to consider how there can be a shared terrain of culture. While *Trinity Cube* was not intended for a masked palm civet audience, this encounter underscores the importance of thinking sensitively about the ways in which different forms and senses can act as a cultural bridge across these species. This encounter asks us to consider what future forms of shared interspecies infrastructure, cultural and otherwise, can be shared.

What types of artwork can form a cultural bridge between species? This was experimented with elsewhere in the world, in particular, the outdoor installation undertaken by artist Rosella Biscotti in 2016 as part of the Sonsbeek Biennale in Arnham, Netherlands, curated by ruangrupa. Biscotti, in developing the new commission, was considering how the large-scale woods of the part that was deserted by humans at night, likely housed a more-than-human community which largely lived a separate life. The installation *Yow-wow-wow-wow-wow* (2016) consisted of bricks, sticks, twine, and lettuce, referencing the Italian art movement from the 1960s and 1970s that utilised such simple materials as these. Biscotti created sculptures with these bricks, suspended balls of sticks and organic matter between trees, and in a low brick-lined circle placed a red ball from a nearby zoo that lions would play with in their cage. In her written statement, Biscotti highlighted how the installation was intended for human visitors during the day and then more-than-human visitors in the evening. Although she did not know who this more-than-human community was, she created the installation as a gesture or experiment for these two audiences. Importantly for her nocturnal audience, there were no cameras in place; rather, she respected their right to opacity. Theorist Edouard Glissant (2010, p. 190), when writing about the meeting and relationship between human cultures, calls for the "right to opacity". This opaqueness creates a more equal footing between cultures as they can mediate their own exchange without being already completely known or even assimilated by other cultures. Glissant sees this opacity as key to a foundation for freedom as well as the beginnings of a respect that can foster a real relation. This right to opacity can be seen as also a critical foundation for interspecies cohabitation. In this case, we are invited to reflect on the nocturnal world of the more-than-human inhabitants who have changed their habits to exist out of sight of the daytime human visitors to the area. While Biscotti (2025) respects this separation, it was the red ball in the circle of bricks that became a bridge in the work. The artist would find many small footprints around the ball in the morning, and it was evident that the ball was quite a popular plaything for the night-time community. This simple gesture of creating an installation in a space shared by humans and more-than-humans began with the acknowledgement of these two communities and by attempting to find entry points for both communities to the artwork—if they were interested.

While *Don't Follow the Wind* has deployed cameras and microphones on the site, as humans cannot freely access the farm or live there, there is also a sensitivity by the collective about what these images are for. Thinking towards to the future arrival of human visitors and the future inhabitants in the area, the moving image surveys help to show which creatures are using which areas of the farm, and perhaps give an insight into their routine and the key points on the farm for the more-than-human community, which otherwise might be overlooked by humans. This survey allows for a future potential shift of the exhibition to accommodate both human and more-than-human guests. This helps the collective and artists to see how different communities are already using the space and what can be done in the future to accommodate a more long-term cohabitation on the site.

As humans will be potentially moving back into their homes by the end of 2030, after almost two decades of displacement, this return—albeit likely in smaller numbers—provides an opening for the returning and new residents to form a different relation with the more-than-human inhabitants. Trevor Paglen's *Trinity Cube* and *Don't Follow the Wind* more broadly have begun a process whereby a future interspecies culture can be developed and appreciated. Learning about different human cultures has long been promoted as a way of reducing conflict—if you know your neighbour, you are less likely to harm them. As such, this form of fostering connections through an exhibition can provide a testing ground for this new type of interspecies structures. These structures might have specific local affects softening the relationship between the human and more-than-human population. More broadly, this sensitive approach to re-learning one's environment, taking careful and considerate approaches in navigating this environment, can help to lay the ontological ground rules for a broader renegotiation of how humans can co-exist with others in their shared environment.

References

Biscotti, Rosella. 8 February 2025. Conversation with the author.

Dombrovski, Valery, Zhurauliou, Dzmitry and Ashton-Butt, Adham. November 2022. 'Long-Term Effects of Rewilding on Species Composition: 22 Years of Raptor Monitoring in the Chernobyl Exclusion Zone'. *Restoration Ecology*. 30, no. 8. Available at: https://doi.org/10.1111/rec.13633 [Accessed 7/25].

Glissant, Édouard. 2010. *The Poetics of Relation*. Ann Arbor, MI: University of Michigan Press.

Hirsch, Nikolaus and Waite, Jason. 2021. *Don't Follow the Wind*. Critical Spatial Practice 12. Berlin: Sternberg Press.

Kohso, Sabu. 2020. *Radiation and Revolution*. Durham: Duke University Press.

Masuda, Ryuichi et al. June 2010. 'Origins and Founder Effects on the Japanese Masked Palm Civet Paguma Larvata (Viverridae, Carnivora), Revealed from a Comparison with Its Molecular Phylogeography in Taiwan'. *Zoological Science*. 27, no. 6. Available at: https://doi.org/10.2108/zsj.27.499 [Accessed 7/25].

Nixon, Rob., 2011. *Slow Violence and the Environmentalism of the Poor*. Cambridge: Harvard University Press.

Saito, Rie et al., December 30, 2021. 'Monitoring of Radioactive Cesium in Wild Boars Captured Inside the Difficult-To-Return Zone in Fukushima Prefecture Over a Five-Year Period'. Available at: https://doi.org/10.21203/rs.3.rs-1181824/v1 [Accessed 7/25].

Szczygielska, Marianna, Kuen, Laura, O'Mahony, Kieran and Keil, Paul G. 2022. 'Feral Ethics: Thinking Feral with Pigs'. Available at: https://doi.org/10.13140/RG.2.2.19094.22081 [Accessed 7/25].

Conclusion

The Remainderless Destruction of Immaterial Objects – Nuclear Apocalypse and the Digital Archive

Min-Kyoo Kim

Radiant Objects

In a dystopian vision of the nuclear wasteland in James Cameron's *The Terminator* (1984), what remains of humanity lives underground, sheltering from the autonomous machines hellbent on driving them to extinction. Within these tunnels, a soldier of the Resistance, Kyle Reese, unfolds a faded Polaroid picture, a peacetime image of Sarah Connor – for whom he will go back in time, to protect her from the Terminator. Beyond mere sentimentality, however, this Polaroid picture hands Kyle a strategic advantage over the Terminator; the archives were mostly destroyed in the nuclear apocalypse, so the machines cannot immediately identify Sarah, affording Kyle precious time to find her. I read this importance of Sarah's photograph in *The Terminator*'s time-travelling narrative alongside Roland Barthes's infamous and intimate reflection on an image of his deceased mother in *Camera Lucida*, the auspicious words of which are germane to this particular collection: within the photograph, Barthes claims, lies a trace of the "*radiant*, irreducible core" of those we love, a trace that may enable us to even revisit the past (2000, p. 75, emphasis added).

I open with this radiant object of Sarah's photograph, an example from science fiction, to illustrate how the archive not only protects the past, but also – potentially – our present and future. This is the central insight of Jacques Derrida's Nuclear Criticism, as presented in his 1984 *Diacritics* article, "No Apocalypse, Not Now". Amid heightening fears of Mutual Assured Destruction in the 1980s – the same pretext that also informs the nuclear anxieties of *The Terminator* – Derrida foresaw the potential "total and remainderless destruction of the archive" (1984, p. 27). Crucially, this unprecedented destruction of the archive would foreclose the movement of what Derrida terms as "*survivance*", the capacity to signify and commemorate death (1984, p. 28). For Derrida, this *survivance* is the very function of the archive, assuring us of the future: that is to say, the archive yields the hope that when we die, we too may be remembered.

Although Derrida's Nuclear Criticism focuses primarily on language and literature, I extend his analysis of *survivance* to objects beyond just books,

DOI: 10.4324/9781003622185-11

to consider photographs, films and other ephemera. It is this diffused understanding of *survivance* that emerges in the edited collection, as conceptualised in Atsuhide Ito's contribution. For Ito, the 'radiant objects' of the Hiroshima Peace Memorial Museum – namely, Shoji Yoshioka's glasses or Osamu Nanko's leather belt – offer mediums of return. As Ito clarifies, the return to these victims is not a regressive, fetishist one. Instead, it is an ethical gesture that obliges us to recognise the reverberating afterlives of nuclear violence, to work for justice in the present and to protect future generations from such a fate. Hence the connection with Derrida's *survivance*; we see how commemorating the past, and encountering those no longer with us, actually sits "at the very heart of life" itself (1984, p. 28).

Thus arises one of the ultimate ironies of the Atomic Age, as Jessica Holtaway signals in her introduction to this collection: that nuclear proliferation, be it in the civil or military sphere, has inspired a seemingly endless production of radiant objects to be collected, studied or even bought in souvenir shops, as in Bridget Kennedy, Robin Tarbet and Helen Grove-White's discussion of the rainbow-striped rubber eraser from the Wylfa Nuclear Power Plant. Perhaps Daniel Beck's chapter exemplifies this irony through his exploration of a former nuclear weapons depository in Warwickshire, UK, which now houses the contents of the British Film Institute (BFI)'s National Film and Television Archive. Beck's attention to this infrastructural entanglement illustrates the fraught co-existence of the nuclear and the archive; as Derrida later expands on in *Archive Fever*, the archive is in fact "made possible by the death, aggression and destruction-drive" (1996, p. 94). Applying this dynamic to the context of the Atomic Age, it is as though the nuclear inspires the very practices of documenting and archiving that it simultaneously threatens to irrevocably destroy.

This irony is also evoked in Yoi Kawakubo's contribution, a poignant reflection on his artistic practice inside the Fukushima exclusion zone. Kawakubo's initial reluctance to produce any such radiant objects indicates the ethical constraints of documenting the nuclear, with the artist noting his anxieties of a possible "amoral complicity" in representing this catastrophe. Kawakubo's frank admission echoes that of the photographer Yoshito Matsushige, who took some of the earliest known images of the immediate aftermath of the atomic bombing of Hiroshima. Recounting his discovery of several corpses in a burnt trolley car, Matsushige stated in an interview: "I thought about taking a picture. I even put my hand on the camera. But it was so hideous that I couldn't do it" (Cook and Cook 1992, p. 394). In their respective accounts, then, Kawakubo and Matsushige both attest to the ethical complexities of documenting the nuclear event, a trauma that appears to simultaneously invite and resist representation. It is this double-movement of the nuclear that Abé Mark Nornes, in his survey of films about the atomic bombings of 1945, has influentially termed the "specter of impossibility" (1996, p. 122).

What Kawakubo did eventually produce in the Fukushima exclusion zone further testifies to this spectre of impossibility, as that which moves both in and beyond the frame of visual representation. As it happens, I had the opportunity to see one of Kawakubo's photographic prints, *If the Radiance of a Thousand Suns Were to Burst into the Skies at Once V* (2019), at the "Picturing the Invisible" exhibition at the Heong Gallery, Cambridge, in 2023. Witnessing the effect of radioactive exposure on the silver halide film was breathtaking. Or rather, I became all too aware *of* my own breathing in this unsettling encounter with a vibrant trace of the nuclear disaster in Fukushima. But beyond just this visceral force, what we also observe in the scratches and abrasions of Kawakubo's prints, I argue, articulates the complex relationship between the archive and the nuclear, wherein the latter simultaneously inspires but also threatens the practice of the former. Kawakubo's prints thus evince how the nuclear creates radiant objects precisely through its distorting, deleterious effects. The nuclear creates, in other words, at the same time as it destroys. Susan Schuppli – a key interlocutor in both Kawakubo and Beck's chapters – expands on this double-movement through the concept of the 'material witness', by which objects exposed to radioactivity are perceived not just as the *evidence* of nuclear destruction, but as bona fide *events* of them also (2020, p. 3).

Therefore, what emerges through this collection's understanding of radiant objects is a series of inherent contradictions in their ontology. Radiant objects offer returns, but perhaps they deliver us only to the limit of our comprehension. Moreover, radiant objects are created by the same force that threatens to destroy not only them but also potentially us, the beholder. Indeed, as Tim Andreae soberly states in his reflections on the radioactive waste from the Idaho National Laboratory, the toxicity of nuclear materials is likely to outlast any container that we can build to hold them. To adapt Marshall McLuhan's oft-cited phrase, the *failure* of the medium is the message here. Radiant objects can thus be read through Schuppli's framework as both forms of evidence and as unfolding events themselves. And so, as Holtaway states in her introduction, the radiant object can be summarised as that which simultaneously enlightens yet eludes.

But in this coda to the collection, I also wish to extend these insights into the contradictions of the radiant object. The majority of the contributions here naturally focus on objects with a concrete materiality – such as the aforementioned rubber eraser, the anti-nuclear pamphlets in Marc Herbst's chapter, or Michael Sanders's 'contaminated tartans'. Each object is entangled, intentionally or inadvertently, with the otherwise indeterminable or incommensurable character of radioactivity. Yet in themselves, the radiant objects remain in the realm of materiality, which can be displayed in an exhibition or indeed even handled and bought. Seeking to unsettle the privileged focus of the material object, I conclude the collection by reflecting on the increasingly digital – and seemingly *immaterial* – basis of the archive today. For it is this digitisation of the archive, the digitisation of radiant objects, that may result in the most poignant irony of all: the fact that the attempt to preserve through digitisation may bring the archive ever closer to the remainderless destruction that Derrida foresaw decades ago.

Digital Archive

In summer 2023, Meta launched its new social media app, Threads. A BBC News article (Dale 2023) remarked upon how, in Britain, the name of this service conjured up memories of an infamous BBC film aired nearly 40 years prior – Mick Jackson and Barry Hines's harrowing dramatisation of nuclear apocalypse, *Threads* (1984). The film depicts the total collapse, or unthreading, of British society in the aftermath of a nuclear exchange between the Cold War superpowers. Those who initially survive this conflict subsist in a nuclear winter. In the film's prophetic and disturbing vision of Derrida's Nuclear Criticism, books – these last vestiges of the peacetime archive, these radiant objects that document and preserve the past – completely disappear. And so, David Forrest and Sue Vice (2017, p. 151) observe that without adequate structures to support it, language itself slowly becomes undone, replaced only by an "alien sense of unfeeling". Future generations are seen, or rather heard, communicating through barely intelligible or emotive sounds. Even the film itself concludes on an abrupt, mute freeze-frame, as though this very work of cultural representation before us has also been corrupted.

I evoke this uncanny coincidence of *Threads* to once again illustrate the close and ironic relationship between the archive and nuclear apocalypse – namely, that a worldwide social media network might also summon one of the most terrifying visions of an entirely disconnected, desolate nuclear wasteland. Yet this is precisely the irony encoded into the Internet and the very concept of digital communications and archives. As Alexander Galloway (2004, p. 29) asserts in *Protocol: How Control Exists after Decentralization*, "while many have debated the origins of the Internet, it's clear that in many ways it was built to withstand nuclear attack". During the 1960s, a RAND Corporation researcher, Paul Baran, advocated for digital communications to enable the American military to maintain its critical command-and-control networks in the aftermath of a potentially catastrophic nuclear attack on its centralized switching facilities (Gartzke and Lindsay 2018, p. 198). Baran's research laid the technical foundations for what became the ARPANET – the early, experimental precursor to the modern-day Internet. Reflecting on this lineage, Bradley Fest (2013, p. 83) stresses that cybernetic and nuclear technologies must be read as "structurally and historically intertwined", with the shared nomenclature of *Threads*/Threads being only the latest iteration of this decades-long entanglement.

Since its initial conception amid the nuclear anxieties of the Cold War, the Internet has become something of a hyperobject, as alluded to in Holtaway's introduction. At once unimaginable in terms of its vastness, yet viscous in the way that phones and computers have become indispensable appendages of our bodies, the Internet does more than merely facilitate communication. Instead, as any reader will already know, its digital expanses are increasingly the domain of our archives, both public and private. For example, in July 2025, the Internet Archive – a nonprofit website providing access to various forms of digitised media, from books to films – was granted federal depository status in the US. This now means that the Internet Archive, which is entirely online, will also be tasked with helping to digitise and store federal government publications, as part of the website's broader objective of enabling "Universal Access to All Knowledge". This vision of the Internet Archive is admirable while also being truly utopian in the etymology that Kawakubo reminds us of, with the term being derived from the Latin U-topos, literally "no-place". It is this very utopian character of the universal, digital archive – that inhabits the clouds in a kind of groundless existence – that leads Fest (2013, p. 84) to warn that the attempt to digitise all knowledge, this "contemporary archival impulse to collect, store and document everything, has the potential to be eschatologically threatening".

Fest's caution is far from alarmist or dystopian. In autumn 2024, a cyberattack rendered the Internet Archive – with its estimated collection of 44 million books and texts, over 10 million videos and nearly 5 million images – temporarily inaccessible to users. Moreover, this incident echoed a similar attack, albeit with more significant consequences, at the British Library in October 2023 – an event that has been described as "one of the worst cyber attacks in British history" (quoted in Ash 2024). In the weeks following this cyberattack, which crippled the British Library's digital infrastructure, there was only minimal access to the vast archive of 170 million items. This prompted *The New Yorker* (Knight 2023) to report, in no uncertain terms, that "the nation's primary wellspring of knowledge was out of commission". Even at time of writing this chapter, in July 2025, the Library is still undertaking its recovery efforts: for example, the Electronic Legal Deposit, the digital archive of all publications in the UK and Ireland, continues to operate with only limited functionality, with no definitive timeline established for a complete restoration of this service. The subsequent and ongoing impact on research, as the writers in this collection would undoubtedly attest, has been significant. But most worrisome of all, perhaps, is not just what has already *happened*, but how these cyberattacks have glimpsed at the future prospect of an altogether more permanent and irreversible erasure of the digital archive.

How exactly, then, does this precarity of the digital archive relate to a broader reflection on the remainderless destruction of nuclear apocalypse? In response, I return to Daniel Beck's chapter here, in which he concludes that although the BFI archive may be secured through processes such as digitisation, there remains value in the preservation of physical media – in other words, of maintaining radiant objects as they are, with their materiality bearing the inevitable traces of the passage of time. This may, in the example of Dimitra Gkitsa's case study of former nuclear bunkers in Albania, even open up these radiant objects to subsequent re-interpretation or indeed, misinterpretation. That being said, I still agree with Beck's privileging of the physical format, albeit with an added caveat. For, as I have sought to outline in the aforementioned cases of the Internet Archive and the British Library, it is not necessarily true that the processes of digitisation secure the archive. On the contrary, I have traced the intertwined development of digital and nuclear technologies to argue that today, our modern and increasingly *immaterial* archives are more vulnerable to the threat of remainderless destruction than when Derrida initially foresaw this threat during the Cold War.

This prospect of a remainderless destruction of the archive is intimated in Annie Jacobsen's *Nuclear War: A Scenario* (2024), a detailed and terrifying minute-by-minute breakdown of a hypothetical nuclear conflict between the US and North Korea. In a situation that Jacobsen (2024, p. 264) refers to as "Electric Armageddon", she draws attention to the electromagnetic pulse (EMP) of a nuclear weapon detonated in the ionosphere. At such a high altitude, the EMP would be silent and pose no direct threat to human or animal life. Yet should it occur, America's 11,000 utility-scale electric power stations, 22,000 generators, and 55,000 substations would experience catastrophic failure. Critical infrastructure across the nation would almost immediately shut down:

> (n)o streetlights, no tunnel lights, no lights at all, only candles, until there are none left to burn. No gas pumps, no fuel. No ATMs. No cash withdrawals. No access to money. No cell phones. No landlines. No calling 911. No calls at all. No emergency communication systems except some high-frequency (HF) radios. No ambulance services. No hospital equipment that works.
>
> (Jacobsen 2024, p. 267)

And although Jacobsen herself does not list the Internet, or the data centres that sustain all our digital archives, other commentators have recently speculated upon the ruinous, and perhaps irrecuperable, effects that an EMP would also wreak on such infrastructure (e.g., Malekos Smith 2024). Through its imperceptible shockwaves, the EMP of a nuclear weapon would precipitate, in Derrida's (1984, p. 27) words, the possibly total and remainderless destruction of the archive – an unprecedented destruction that would surpass any such archival crisis before it, including any "burning of a library, even that of Alexandria". We would be plunged, therefore, into a literal and digital dark age.

The ultimate irony from all this – or at this stage, maybe the appropriate term would be tragedy – is that the utopian dream of endless archival accumulation, alongside the desire for an archive liberated from the material constraints of time and space, is exactly what renders the archive more precarious to the prospect of its remainderless destruction. I thus agree with Fest's (2013,

p. 96) forceful conclusion that "(w)e must be aware of the relative inhumanity that drives and results from the hyperarchival impulse". Now, it may be objected that in the context of an 'Electric Armageddon' – unfolding alongside the devastation of a full-blown nuclear war – it seems almost irrelevant to worry about the imperilment of the archive, when human existence itself appears at stake. But here, I ask the reader to stay with this particular strand of trouble. Neither Jacobsen's book nor Derrida's Nuclear Criticism rules out the possibility that some humans, in some shape or form, would survive this initial onset of nuclear apocalypse. Perhaps this illustrates either the inability, or indeed unwillingness, of anthropocentric thinking to reach beyond the immediate horizon of our own species – thus further exposing the epistemological complexities of grasping the 'more-than-human' imaginaries that Jason Waite so eloquently indicates in his contribution to the collection. While cognizant of this anthropocentric bias, and maybe even being guilty of it, I too stay with Jacobsen and Derrida's tentative prediction that some form of humanity would live on. I inhabit this scenario, not out of blind optimism, but instead to confront the question of what it means to live in a world without the archive, and without the structures of language and communication that the archive enables. I am then reminded of the concluding scenes of *Threads*, in which the 'schools' of the future rely upon warped videotapes played on televisions hanging precariously by worn extension cords. I hear the garbled words of these young people, now barely able to communicate themselves. Barely able, as it transpires, even to mutter goodbye to their dying loved ones. This is the end of *survivance* as Derrida first foresaw in his Nuclear Criticism – and with this cessation of *survivance,* I posit, follows the concluding moments of what we may assuredly call humanity. This returns us to my opening statement of the paragraph, which has hitherto guided the chapter: the ultimate irony of the Atomic Age is that the steps we have taken to prepare ourselves and the archive *for* a possible apocalypse may, in fact, *have prepared the apocalypse itself.*

Conclusion

To conclude the collection, I return to my opening with *The Terminator*, animating this chapter's own kind of cyclical time-loop. Except this time, I re-imagine (as a seemingly interminable list of sequels has been prone to do) James Cameron's original text, to consider what was to happen if there was no Polaroid photo taken of Sarah Connor, this radiant object that guides Kyle back to the past, and in so doing, enables him to save the future. Or alternatively, maybe the only image of Sarah was digital, and stored on a cloud drive that dissipated in the aftermath of a catastrophic EMP. I invoke these revised scenarios to imagine how, without the radiant objects that anchor our relationship to others – be they humans, non-humans or even other inanimate objects – humanity would be left adrift, unmoored without any sense of its past or security in its future.

What can we do, then, with this awareness of an impending archival crisis? Fest (2013, p. 95) helpfully notes – to indulge one, final etymological exercise in this collection – that the term 'crisis' is derived from the Ancient Greek word *krisis*, meaning 'decision'. In the opening moments of the chapter, I reflected on the value, however contested or evanescent, of the radiant objects accrued in our archive of the Atomic Age. This inventory of objects, as Holtaway puts it, obliges us to consider past and present impacts of the nuclear on exposed populations, even on those we may not be able to name or easily visualise, and yet for whom we still maintain a responsibility. The crisis that faces us now is how best to protect these radiant objects, and how to hold strong and true to the connections that these objects proffer. In this light, I have sought to argue that the digitisation of the archive, this translation of radiant objects into immateriality, may bring the archive ever closer to its remainderless destruction. Confronted with this fallacy, engineered through the historic entanglement of civil and military technologies, we must decide how to preserve the archive and its manifold radiant objects. These questions of how we attend to and preserve these traces and voices of the past may just determine our future.

References

Ash, Lamorna. 2024. 'Thanks to a shadowy hacker group, the British Library is still on its knees. Is there any way to stop them?' *The Guardian*, February 6. Available at: https://www.theguardian.com/commentisfree/2024/feb/06/hacker-british-library-cybersecurity-cybercrime-uk [Accessed 7/25].

Barthes, Roland. 2000. *Camera Lucida*. London: Vintage.

Cook, Haruko Taya and Cook, Theodore F. 1992. *Japan at War: An Oral History*. New York: W. W. Norton and Company.

Dale, Tim. 2023. 'Threads: The BBC drama which affected a generation of viewers'. BBC, July 12. Available at: https://www.bbc.com/news/uk-england-south-yorkshire-66122775 [Accessed 7/25].

Derrida, Jacques. 1984. 'No Apocalypse, Not Now (Full Speed Ahead, Seven Missiles, Seven Missives.' *Diacritics* 14 (2), pp. 20–31. Translated by Catherine Porter and Philip Lewis. Available at: https://doi.org/10.2307/464756.

Derrida, Jacques. 1996. *Archive Fever: A Freudian Impression*. Translated by Eric Prenowitz. Chicago: University of Chicago Press.

Fest, Bradley J. 2013. 'Apocalypse Networks: Representing the Nuclear Archive.' In Michael Blouin, Morgan Shipley and Jack Taylor (eds.), *The Silence of Fallout: Nuclear Criticism in a Post-Cold War World*. Newcastle-upon-Tyne: Cambridge Scholars Publishing. pp. 81–103.

Forrest, David and Vice, Sue. 2017. *Barry Hines: Kes, Threads and Beyond*. Manchester: Manchester University Press.

Galloway, Alexander R. 2004. *Protocol: How Control Exists after Decentralization*. Cambridge, Mass.: The MIT Press.

Gartzke, Erik and Lindsay, Jon R. 2018. 'The Cyber Commitment Problem and the Destabilization of Nuclear Deterrence.' In Herbert Lin and Amy Zegart (eds.), *Bytes,*

Bombs, and Spies: The Strategic Dimensions of Offensive Cyber Operations. Washington, D.C: Brookings Institution Press, pp. 195–234.

Jacobsen, Annie. 2024. *Nuclear War: A Scenario*. London: Transworld Publishers Ltd.

Knight, Sam. 2023. 'The disturbing impact of the cyberattack at The British Library.' *The New Yorker*. December 19. Available at: https://www.newyorker.com/news/letter-from-the-uk/the-disturbing-impact-of-the-cyberattack-at-the-british-library [Accessed 7/25].

Malekos Smith, Zhanna L. 2024. 'The specter of EMP weapons in space'. *Carnegie Council for Ethics in International Affairs*, March 27. Available at: https://carnegiecouncil.org/media/article/the-specter-of-emp-weapons-in-space [Accessed 7/25].

Nornes, Abé Mark. 1996. 'The Body at the Center – The Effects of the Atomic Bomb on Hiroshima and Nagasaki.' In Mick Broderick (ed.), *Hibakusha Cinema: Hiroshima, Nagasaki and the Nuclear Image in Japanese Film*. London: Kegan Paul International, pp. 120–159.

Schuppli, Susan. 2020. *Material Witness: Media, Forensics, Evidence*. Cambridge, Mass.: The MIT Press.

Index

Note: Page references in *italics* refer to figures and with "n" refer to endnotes.

For Product Safety Concerns and Information please contact our EU representative GPSR@taylorandfrancis.com Taylor & Francis Verlag GmbH, Kaufingerstraße 24, 80331 München, Germany